LINEAR
ALGEBRA

LINEAR ALGEBRA

Li-Yong Shen, PhD
(University of the Chinese Academy of Sciences)

Haohao Wang, PhD
(Southeast Missouri State University)

Jerzy Wojdylo, PhD
(Southeast Missouri State University)

MERCURY LEARNING AND INFORMATION
Dulles, Virginia
Boston, Massachusetts
New Delhi

Publisher: David Pallai
Mercury Learning and Information
22841 Quicksilver Drive
Dulles, VA 20166
info@merclearning.com
www.merclearning.com
(800) 232-0223

Li-Yong Shen, H. Wang, J. Wojdylo. *Linear Algebra.*
ISBN: 978-1-683923-76-3

Library of Congress Control Number: 2018962730

181920321 This book is printed on acid-free paper in the USA.

CONTENTS

PREFACE

Linear algebra is one of the most widely used mathematical theories and has applications in virtually every area of mathematics, including multivariate calculus, differential equations, and probability theory. The purpose of this book is to bridge the gap between the abstract theoretical aspects and the computational applications of linear algebra. The aim of this book is two-fold: to introduce the fundamental concepts of linear algebra and to apply the theorems in computation-oriented applications. There are many good introductory texts on linear algebra, and the intention of this book is to be a supplement to those texts, or to serve as a text for senior undergraduate students or first year graduate students, whose interests are computational mathematics, science, engineering, and computer science. The presentation of the material combines definitions and proofs with an emphasis on computational applications. We provide examples that illustrate the use of software packages such as *Mathematica*, *Maple*, and *Sage*.

This book has evolved from our experience over several years of teaching linear algebra to mixed audiences of upper division mathematics majors, beginning graduate students, and students from the fields of science and engineering that rely heavily on mathematical methods. Our goal in writing this book has been to develop a text that addresses the exceptional diversity of the audience, and introduce some of the most essential topics about the subject of linear algebra to these groups. To accomplish our goal, we have selected material that covers both the theory and applications, while emphasizing topics useful in other disciplines.

Throughout the text, we present a brief introduction to some aspects of abstract algebra that relate directly to linear algebra, such as groups, rings, modules, fields and polynomials over fields. In particular, the last section of this book is dedicated to the matrix decomposition over principle ideal domains, because this structure theorem is a generalization of the fundamental theorem of finitely-generated abelian groups, and this result provides a simple framework to understand various canonical form results for square matrices over fields.

We use the material from the book to teach our own elective linear algebra course, and some of the solutions to the exercises are provided by our students. It is our hope that this book will help a general reader appreciate the abstract mathematics behind the real applications.

By design, each chapter consists of two parts: the theoretical background and the applications, which make the book suitable for a one semester course in linear algebra that can be used in a variety of contexts. For an audience composed primarily of mathematics undergraduate majors, the material on the theories of abstract vector spaces, linear transformations, linear operators, orthogonal bases, and decomposition over rings can be covered in depth. For an applied mathematics course with students from the fields of science and engineering that rely heavily on mathematical methods, the material on applications of these areas such as linear codes, affine or projective transformations, geometry of transformations, matrix in graph theory, image processing, and QR decomposition, can be treated with more emphasis. In the applications, we allow ourselves to present a number of results from a wide range of sources, and sometimes without detailed proofs. The applications portion of the chapter is suitable for a reader who knows some linear algebra and a particular related area such as coding theory, geometric modeling, or graph theory. Some of the applications can serve as a guide to some interesting research topics.

The prerequisite for this book is a standard first year undergraduate course in linear algebra. In Chapter 1 and Chapter 2 we start with a quick review of the fundamental concepts of vector spaces and linear transformations. To better understand the behavior of a linear transformation, we discuss the eigenvectors in Chapter 3, where the eigenvectors act as the "axes" along which linear transformations behave simply as stretching, compressing, or flipping, and hopefully make understanding of linear transformations easier. Because one can perform some operations on vectors by performing the same operations on the basis, we study orthogonal bases in Chapter 4. In particular, we study linear transformations relative to orthonormal bases that faithfully preserve the linear properties and the metric properties. Finally, in Chapter 5, we focus on the matrix decomposition over real or complex numbers and over principle ideal domains.

This book should be thought of as an introduction to more advanced texts and research topics. The novelty of this book, we hope, is that the material presented here is a unique combination of the essential theory of linear algebra and computational methods in a variety of applications.

LIST OF FIGURES

CHAPTER 1

VECTOR SPACES

1.1 Vector Spaces

Linear algebra is an area of mathematics which deals with vector spaces and linear mappings between vector spaces. The fundamental subjects studied in linear algebra includes lines, planes, and subspaces. For example, the condition under which a set of n hyperplanes intersect in a single point can be investigated via finding the condition for a system of linear equations to have a unique non-trivial solution. This system of linear equation can be represented by vectors and matrices. In this section, we will focus on the vector spaces.

We will first recall the notation of a *set*. A set is a collection of distinct objects, the objects are called elements of the set. For instance, all integer numbers form a set $\mathbb{Z} = \{..., -2, -1, 0, 1, 2, ...\}$.

A set X is called an empty set, denoted by $\varnothing$, if X has no elements. We denote $x \in X$ if an element x is included in a set X, otherwise denote $x \in X$. For two sets X and Y, we call X a subset of Y (or Y a superset of X), denoted by $X \subseteq Y$ (or $Y \supseteq X$), if all the elements *of* X are included in Y. Note that the empty set is a subset of any set. We say X is a strict (alternatively proper in some references) subset of Y (or Y is a strict superset of X), denoted by $X \subset Y$ (or $Y \supset X$) if $X \subset Y$ and there is an element y such that $y \in$ Y and $y \notin X$.

We give three basic operations for two sets:

1. Union of two sets X and Y is the set of elements which are in X, in Y, or in both X and Y, denoted by $X \cup Y = \{a \mid a \in X \text{ or } a \in Y\}$;

2. Intersection of two sets X and Y is the set of elements which are in both X and Y, denoted by $X \cap Y = \{a \mid a \in X \text{ and } a \in Y\}$;

3. Difference of two sets X and Y is the set of elements in X but not in Y, denoted by $X \backslash Y = \{a \mid a \in X \text{ and } a \notin Y\}$.

Definition 1.1.1

A **ring** R consists of a set with two operations $(+, *) : R \times R \rightarrow R$ with the properties:

1. Addition is commutative and associative.

2. There is a unique element $0 \in$ R, such that $x + 0 = x$ for all $x \in R$.

3. To each $x \in R$, there corresponds a unique element $-x \in R$ such that $x + (-x) = 0$.

4. Multiplication is associative.

5. Multiplication is distributive over addition on the left (and/or right).

If the multiplication is commutative, then R is called a **commutative ring.** If there is a unique non-zero element $1 \in R$ such that $x * 1 = 1 * x = x$ for all $x \in R$, then the ring is called a **ring with identity**. If for each non-zero element $x \in R$, there corresponds a unique element $y \in R$ such that $x * y = y * x = 1$, then we say R is a **division ring.**

For a ring R, if R satisfies the first three conditions, then $(R, +)$ is a **commutative** *or* **abelian additive group**. When a ring R is with unity, then $(R, *)$ is a commutative multiplicative group.

Basically, a *field* is an algebraic structure in which every linear equation in a single variable can be solved.

Definition 1.1.2

Let $\mathbb{F}$ be a set with two operations $(+, *) : \mathbb{F} \times \mathbb{F} \rightarrow \mathbb{F}$ with the following properties, then $\mathbb{F}$ is a **field.**

1. Addition is commutative and associative.

2. There is a unique element $0 \in \mathbb{F}$, *such that* $x + 0 = x$ for all $x \in \mathbb{F}$.

3. To each x $\in \mathbb{F}$, there corresponds a unique element $-x \in \mathbb{F}$ such that $x + (-x) = 0$.

4. Multiplication is commutative and associative.

5. There is a unique non-0 element $1 \in \mathbb{F}$ such that $x * 1 = x$ for all $x \in \mathbb{F}$.

6. To each non-zero $x \in \mathbb{F}$, there corresponds a unique element $x^{-1} \in \mathbb{F}$ such that $x * x^{-1} = 1$.

7. Multiplication is distributive over addition.

We recall that the set of rational numbers $\mathbb{Q}$, the set of real numbers $\mathbb{R}$, and the set of complex numbers $\mathbb{C}$, are all fields with the usual addition and multiplication, and $\mathbb{Q} \subset \mathbb{R} \subset \mathbb{C}$.

We know that the difference between rings and fields is that fields are commutative division rings. The main difference between rings and fields is that one can divide in fields, but in general one cannot always divide in rings. For instance, the set of integers is a commutative ring; if we include all the multiplicative inverses of non-zero integers, then we get a field, the field of rational numbers. A polynomial ring is a commutative ring formed from the set of polynomials in one or more indeterminates, or variables, with coefficients in another ring. For example $\mathbb{R}[x]$ is a polynomial ring in one variable with coefficients in $\mathbb{R}$. If we include all the multiplicative inverses of non-zero polynomials, then we get a field, the field of rational functions, $\mathbb{R}(x)$.

One can verify that the set of $n \times n$ matrices over $\mathbb{R}$ is a ring, but not a commutative ring. This ring has identity, i.e., the $n \times n$ identity matrix, but this is not a ring with unity.

A *vector space,* or sometimes called a *linear space,* is a collection of objects called *vectors* that satisfies certain properties.

Definition 1.1.3

A **vector space** or **linear space** V over a field $\mathbb{F}$ consists of a non-empty set V together with two operations $+ : V \times V \to V$, and $* : \mathbb{F} \times V \to V$ which satisfy the following:

1. A rule (or operation) +, called vector addition, which associates with each pair of vectors $\mathbf{a}, \mathbf{b} \in V$ a vector $\mathbf{a} + \mathbf{b} \in V$, called the sum of $\mathbf{a}$, $\mathbf{b}$, in such a way that
   (*a*) addition is commutative
   (*b*) addition is associative
   (*c*) there is a zero vector $\mathbf{0}$ which is additive identity
   (*d*) for each vector $\mathbf{a} \in V$, there is $-\mathbf{a} \in V$, such that $\mathbf{a} + (-a) = \mathbf{0}$.

2. A rule (or operation) $*$, called scalar multiplication, which associates with each $c \in \mathbb{F}$ and a vector $\mathbf{a} \in V$ a vector $c\mathbf{a} \in V$, called the product, such that for all $\mathbf{a}, \mathbf{b} \in V$ and $c_1, c_2 \in \mathbb{F}$
   (*a*) $1 * \mathbf{a} = \mathbf{a}$
   (*b*) $(c_1c_2) * \mathbf{a} = \mathbf{c}_1 * (c_2 * \mathbf{a})$
   (c) $c*(\mathbf{a} + \mathbf{b}) = c * \mathbf{a} + c * \mathbf{b}$
   (*d*) $(c_1 + c_2) * \mathbf{a} = c_1 * \mathbf{a} + c_2 * \mathbf{a}$.

Definition 1.1.4

A subset W of a vector space V is called a **subspace** of V if W itself is a vector space under the addition and scalar multiplication inherited from V.

EXAMPLE 1.1.1

Let $\mathbb{F}[x]$ be the collection of all polynomials in one variable x, i.e., the polynomial ring with one variable with coefficients over the field $\mathbb{F}$. Then $\mathbb{F}[x]$ with the usual polynomial addition and multiplication by constants is a vector space over $\mathbb{F}$. Polynomials of degree n are completely determined, by the coefficient of x^k for $k = 0, \ldots, n$. To check that is a vector space, one needs to know how addition and scalar multiplication by elements of $\mathbb{F}[x]$ are defined.

A *matrix* is an important object in linear algebra. A matrix is a rectangular array of numbers or expressions, called entries of the matrix, arranged in rows and columns.

EXAMPLE 1.1.2

Let $M_{n \times n}(\mathbb{F})$ be the set of $n \times n$ matrices with entries in $\mathbb{F}$. Then $M_{n \times n}(\mathbb{F})$ is a vector space over $\mathbb{F}$ with the usual addition and scalar multiplication. Moreover, if we let $D_n(\mathbb{F})$ and $U_n(\mathbb{F})$ be the set of diagonal matrices and upper triangular matrices in $M_{n \times n}(\mathbb{F})$, then $D_n(\mathbb{F})$ and $U_n(\mathbb{F})$ are subspaces of $M_{n \times n}(\mathbb{F})$.

EXAMPLE 1.1.3

An $n \times n$ matrix $A = (a_{jk})$ where a_{ij} denotes the entry of the *i-th* row and *j-th* column over the field $\mathbb{C}$ of complex numbers is **Hermitian** or **self-adjoint** if

$a_{jk} = \overline{a}_k j$, for each j, k; the bar denoting complex conjugation.

The set of all Hermitian matrices is not a subspace of the space of all $n \times n$ matrices over $\mathbb{C}$.

EXAMPLE 1.1.4

Let $M_{m\times n}(\mathbb{F}[x])$ be the set of $m \times n$ matrices with entries in $\mathbb{F}[x]$. Then $M_{m\times n}(\mathbb{F}[x])$ is a vector space over $\mathbb{F}[x]$ with the usual addition and scalar multiplication. In fact, $M_{m\times n}(\mathbb{F}[x])$ is isomorphic to $(M_{m\times n}(\mathbb{F}))[x]$. For instance, 3×3 polynomial matrix of degree 2:

$$\begin{pmatrix} 4 & 5x^2 & x \\ 0 & 3x & 2 \\ 3x+2 & 7x^2-11 & 0 \end{pmatrix} = \begin{pmatrix} 4 & 0 & 0 \\ 0 & 0 & 20 \\ 2 & -11 & 0 \end{pmatrix} + \begin{pmatrix} 0 & 0 & 1 \\ 0 & 3 & 0 \\ 3 & 0 & 0 \end{pmatrix} x + \begin{pmatrix} 0 & 5 & 0 \\ 0 & 0 & 0 \\ 0 & 7 & 0 \end{pmatrix} x^2$$

Definition 1.1.5

Suppose that R is a ring and 1 is its multiplicative identity. A **left** R-module M is a set together with two operations $+ : M \times M \to M$ with an abelian group (M, +), and an operation $* : R \times M \to M$ such that for all r, $s \in R$ and x, $y \in M$, we have:

$$r*(x+y) = x*x + r*y$$
$$(r+s)*x = r*x + s*x$$
$$(rs)*x = r*(s*x)$$
$$1*x = x$$

The operation $*$ of the ring on M is called scalar multiplication. The notation RM indicates a left R-module M. A right R-module M or MR is defined similarly, except that the ring acts on the right; i.e., scalar multiplication takes the form $* : M \times R \to M$, and the above axioms are written with scalars r and s on the right of x and y.

If a ring is non-commutative there is a difference between a left and right action. A given left action does not in general define a right action. Over a commutative ring there is no difference between a left and right module.

A module is abstractly similar to a vector space, but it uses a ring to define coefficients instead of the field used for vector spaces. In fact, a vector space is a module over a field, but in general, modules have coefficients in much more general algebraic objects. It is easy to see that Hermitian matrices is not a module over $\mathbb{C}$.

Following are a few theorems which can be verified by the definition directly.

Theorem 1.1.1

A non-empty set $W \subset V$ is a subspace of a vector space V if and only if the following hold:

1. For all $\mathbf{x}, \mathbf{y} \in W$, $\mathbf{x} + \mathbf{y} \in W$;
2. For every $\mathbf{x} \in W$, and scalar a, the vector $\alpha x \in W$.

Definition 1.1.6

If U, W are subspaces of the vector space V, then we define the sum of U, W to be

$$U + W = \{\mathbf{x} + \mathbf{y} \mid \mathbf{x} \in U, \mathbf{y} \in W\}$$

Similarly, if U_i are subspaces of V, then the sum of U_i is

$$U_i + \cdots + U_k = \{x_i + \cdots + x_k \mid x_i \in U_i, \mathrm{i} = 1, \cdots, k\}.$$

Remark: *We note that the union of two vector subspaces is not a vector or subspace.*

Theorem 1.1.2

Suppose U, W are subspaces of the vector space V. Then $U \cap \mathrm{W}$, and $U + W$ are subspaces of V.

Definition 1.1.7

Let $U_1, \ldots, U_k$ be subspaces of a vector space V. We say V is a **direct sum** *of* $U_1, \ldots, U_k$ and write $V = U_1 \oplus U_2 \oplus \cdots \oplus U_k$ *if*

1. Every vector $x \in V$ can be written as $\mathbf{x} = \mathbf{y}_1 + \mathbf{y}_2 + \cdots + \mathbf{y}_k$ with $\mathrm{y}_i \in U_i$;
2. If $\mathbf{y}_i, \mathbf{w}_i \in U_i$, and $\mathbf{y}_1 + \cdots + \mathbf{y}_k = \mathbf{w}_1 + \cdots + \mathbf{w}_k$, then $\mathbf{y}_i = \mathbf{w}_i$ for all $i = 1, \ldots, k$.

There is a similar definition, namely, the **direct product** $\Pi_{i \in I} V_i$ of a family of vector spaces V consists of the set of all tuples $\mathbf{x}_i \in V_i$ with $i \in I$ where addition and scalar multiplication is performed component wise, and I is either a finite or infinite index set. For example, the direct product is the same as Cartesian product. If X and Y are two sets, then $X \times Y$, the Cartesian product of X and Y is a set made up of all ordered pairs of elements of X and Y, i.e., if $x \in X$, $y \in Y$, then $(x, y) \in X \times Y$.

EXAMPLE 1.1.5

Take $V = \mathbb{R}^3$. We can describe all subspaces: $\{0\}$, $\mathbb{R}^3$, all lines and planes through the origin.

Given any two distinct lines U_1 and U_2 through the origin, we can take the plane $U_1 + U_2$ that they span.

Given a plane U through the origin, and a line W through the origin not in that plane, then $U + W = \mathbb{R}^3$. Furthermore, every vector $v \in \mathbb{R}^3$ is a unique sum of a vector of U and another vector in W, that is $\mathbb{R}^3 = U \oplus W$.

Using the definition of direct sum, one can verify the following theorem:

Theorem 1.1.3

Let $U_1,\ldots,U_k$ be subspaces of a vector space V. Let $W_i = U_1 + \cdots + \tilde{U}i + \cdots + U_k$, i.e., sum of all U_j for $j \neq i$. Then $V = U_1 \oplus \cdots \oplus U_i \oplus \cdots \oplus U_k$ if and only if

1. $V = U_1 + \cdots + U_k$;
2. $U_i \cap W_i = \{0\}$ where $\mathbf{0} \in V$ for each i.

EXAMPLE 1.1.6

Let $\mathcal{W} = \{W_i \mid i \in I\}$ be a collection of subspaces of V where the index set I can be finite or infinite. It is clear that $\cap|_{i\in I} W_i$ is a subspace of V. If I is a finite set, then the set of all finite sums of the vectors from $\Sigma_{i\in I}W_i$ is also a subspace of V, and denote this as $\Sigma_{i\in I}W_i$. If I is an infinite set, let

$$\sum_{i\in I} W_i = \left\{\sum_{i\in I} \mathbf{w}_i \mid \mathbf{w}_i \in \mathbf{w}_i, \forall i \in I\right\},$$

where $\Sigma_{i\in I}\mathbf{w}_i$ means that all $\mathbf{w}_i = 0$ except possibly for finitely many $i \in I$. Therefore, regardless whether I is a finite or infinite set, with the above notation $\Sigma_{i\in I}W_i$ is a subspace of V.

Theorem 1.1.4

Let $\mathbb{F}$ be an infinite field and V be a vector space over $\mathbb{F}$. Then V cannot be the union of a finite number of proper subspaces (subspaces that are strictly contained in V).

Proof: We will prove the claim by contradiction. Let $W_i \subseteq V$ *for* $i = 1,\ldots,n$ be proper subspaces of V. Suppose $V = \bigcup_{i=1}^{n} W_i$. Without loss of generality, we assume that $W_1 \not\subset \bigcup_{i=2}^{n} W_i$.

Let $\mathbf{w} \in W1$ $\mathbf{w} \in W_1 \setminus \bigcup_{i=2}^{n} W_i$, and let $\mathbf{v} \in V \setminus W_1$. *Consider the set* $U = \{\mathbf{w} + a\mathbf{v} \mid a \in \mathbb{F}\}$. Since $\mathbb{F}$ is infinite, *U is infinite. Hence* $U \cap W_k$ is infinite for some index k where $k = 1, 2,\ldots, n$.

Suppose $k \neq 1$. If $(\mathbf{w} + a\mathbf{v})$, $(\mathbf{w} + b\mathbf{v}) \in U \cap W_k$ for some distinct $a, b \in \mathbb{F}$, then $b(\mathbf{w} + a\mathbf{v}) - a(\mathbf{w} + b\mathbf{v}) = (b - a)\mathbf{w} \in W_k$. *And hence* $\mathbf{w} \in W_k$, *contradicting* the assumption that $\mathbf{w} \in W_1 \setminus \bigcup_{i=2}^{n} W_i$. Thus, $k = 1$. But this is not possible, since $W_k = W_1$ is a subspace, $(\mathbf{w} + a\mathbf{v}) - (\mathbf{w} + b\mathbf{v}) = (a - b)\mathbf{v} \in W_1$, which yields $\mathbf{v} \in W1$. This contradicts $\mathbf{v} \in V \setminus W1$. Therefore, we proved the original claim.

Notice that the above theorem may not be true if $\mathbb{F}$ is a finite field. For example, let $\mathbb{F} = \mathbb{Z}_2$, and the elements of the vector space is coming from the finite field $\mathbb{Z}_2$. Then $V = \{(0, 0), (1, 0), (0, 1), (1, 1)\}$. Let subspaces $V_1 = \{(0, 0), (1, 0)\}$, $V_2 = \{(0, 0), (0, 1)\}$, and $V_3 = \{(0, 0), (1, 1)\}$, then $V = V_1 \cup V_2 \cup V_3$.

1.2 Linear Span and Linear Independence

Definition 1.2.1

A vector $\mathbf{b} \in V$ is said to be a **linear combination** of the vectors $\mathbf{a}_1, \ldots, \mathbf{a}_n$ in V if there exist scalars $c_1, \ldots, c_n$ in $\mathbb{F}$ such that

$$\mathbf{b} = \sum_{i=1}^{n} c_i \mathbf{a}_i.$$

The set of all linear combinations of $\mathbf{a}_1,\ldots,\mathbf{a}_n$ is called the **span** of $\mathbf{a}_1,\ldots,\mathbf{a}_n$, and it is denoted by Span $(\mathbf{a}_1,\ldots,\mathbf{a}_n)$.

If a vector space $V = \text{Span}(\mathbf{a}_1, \ldots,\mathbf{a}_n)$, then we say that $(a_1, \ldots,a_n)$ **spans** V, and $(\mathbf{a}_1, \ldots,\mathbf{a}_n)$ is a **spanning sequence** for V.

A vector space V is **finitely generated** if it is possible to find a finite sequence of vectors $(\mathbf{a}_1, \ldots,\mathbf{a}_n)$ such that $V = \text{Span}(\mathbf{a}_1, \ldots,\mathbf{a}_n)$.

EXAMPLE 1.2.1

The real vector space $\mathbb{R}^3$ has $\{(2, 0, 0), (0, 1, 0), (0, 0, 1)\}$ as a spanning set. $\{(1, 2, 3), (0, 1, 2), (-1, 0.5, 3), (1, 1, 1)\}$ is also a spanning set for $\mathbb{R}^3$.

But, the set $\{(1, 0, 0), (0, 1, 0), (1, 1, 0)\}$ is not a spanning set of $\mathbb{R}^3$, since $(0, 0, 1) \in \mathbb{R}^3$ is not in the span of this set. The span of this set is the space of all vectors in $\mathbb{R}^3$ whose last component is zero.

EXAMPLE 1.2.2

Let $A = (\mathbf{a}_1,\ldots,\mathbf{a}_m)$, and $B = (\mathbf{b}_1,\ldots,\mathbf{b}_m)$ be two spanning sequences of a vector space V. Define the union of the two sequences A and B by $A \cup B = (\mathbf{a}_1,\ldots,\mathbf{a}_m, \mathbf{b}_1,\ldots,\mathbf{b}_n)$. It is easy to check that Span(A) ∪ Span(B) ⊂ Span(A ∪ B).

Theorem 1.2.1

Let $A = (\mathbf{a}_1,\ldots,\mathbf{a}_m)$ be a sequence such that $\mathbf{a}_i \in$ V. Then

1. Span(A) is a subspace of V;
2. If W is a subspace of V, and $A \subset W$, then Span(A) $\subset W$.

Proof: One can check the claim by using the definition of span. The second claim says that Span(A) is the "smallest" subspace of V which contains A. If W contains A, and $W \subseteq C$ Span(A), then $W =$ Span(A).

Theorem 1.2.2

Let $A = (\mathbf{a}_1,\ldots,\mathbf{a}_m)$ be a sequence such that $\mathbf{a}_i \in V$ are distinct vectors. If there exists a vector $\mathbf{a}_i$ for some i such that $\mathbf{a}_i = \sum_{j=1 j\neq i}^{m} c_j\mathbf{a}_j$, then Span($A$) = Span($A \setminus \{\mathbf{a}_i\}$).

Proof: It is easy to see that Span(A) ⊇ Span($A \setminus \{\mathbf{a}_i\}$). To see Span($A$) ⊆ Span($A \setminus \{\mathbf{a}_i\}$), let $\mathbf{b} \in$ Span(A), then

$$b = \sum_{j=1}^{m} a_j\mathbf{a}_j, \quad a_j \in \mathbb{F},$$

$$= \sum_{j=1}^{i-1} a_j\mathbf{a}_j + a_i\left(\sum_{j=1, j\neq i}^{m} c_j\mathbf{a}_j\right) + \sum_{j=i+1}^{m} a_j\mathbf{a}_j$$

$$= \sum_{j=1}^{i-1}(a_j + a_i c_j)\mathbf{a}_j + \sum_{j=i+1}^{m}(a_j + a_i c_j)\mathbf{a}_j = \sum_{j=1, j\neq i}^{m}(a_j + a_i c_j)\mathbf{a}_j$$

$$\in \text{Span}(A \setminus \{\mathbf{a}_i\})$$

Thus, Span(A) = Span($A \setminus \{\mathbf{a}_i\}$).

So far, we have seen that any finite subset A of a vector space V determines a subspace Span(A) *of* V. If we let $\mathcal{A}$ *denote the set of all subsets of* V, and $\mathcal{W}$ denote the set of all subspaces of V, then there is a function Span : $\mathcal{A} \to \mathcal{W}$ which sends a subset $A \in \mathcal{A}$ *to* Span(A) $\in \mathcal{W}$. One can show that Span is a function with the following properties:

1. If $A_1 \subseteq A_2 \in \mathcal{A}$, then $\text{Span}(A_1) \subseteq \text{Span}(A_2) \in \mathcal{W}$.
2. If $\mathbf{w} \in \text{Span}(A)$, then there exists a subset $A' \subseteq A$ such that $\mathbf{w} \in \text{Span}(A')$.
3. $A \subseteq \text{Span}(A)$ for all $A \in \mathcal{A}$.
4. For every $A \in \mathcal{A}$, $\text{Span}(\text{Span}(A)) = \text{Span}(A)$.
5. Let $\mathbf{v}, \mathbf{w} \in V$, if $\mathbf{v} \in \text{Span}(A \cup \{\mathbf{w}\}) \backslash \text{Span}(A)$, then $\mathbf{w} \in \text{Span}(A \cup \{\mathbf{v}\})$.

We observe that if a certain element of a spanning set can be spanned by the other elements of the spanning set, then one can reduce the number of elements in the spanning set. This introduces the following concept.

Definition 1.2.2

A finite sequence of vectors $\mathbf{a}_1,\ldots,\mathbf{a}_n$ from a vector space V is **linearly dependent** if there are scalars $c_1, \ldots, c_n$, not all zero, such that $\sum_{i=1}^{n} c_i\mathbf{a}_i = \mathbf{0}$. The sequence is said to be **linearly independent** if

$$\sum_{i=1}^{n} c_i\mathbf{a}_i = \mathbf{0} \quad \Rightarrow \quad c_1 = \cdots c_n = 0.$$

An infinite set of vectors is linearly dependent if it contains a finite subset that is linearly dependent. Otherwise, this infinite set of vectors is called linearly independent.

It is easy to check that if a spanning set contains repeated vectors, or if one of the vectors is a linear combination of the other vectors, then the spanning set is linearly dependent.

Theorem 1.2.3

Let $A - (\mathbf{a}_1,\ldots,\mathbf{a}_n)$ be a linearly independent sequence of vectors of a vector space V. Then

1. If $\mathbf{b} \notin \text{Span}(A)$, then $A \cup \{\mathbf{b}\} - \{\mathbf{a}_1,\ldots, \mathbf{a}_n, \mathbf{b}\}$ is a linearly independent set.
2. If $\mathbf{x} \in \text{Span}(A)$, then $\mathbf{x}$ can be uniquely expressed as

$$\mathbf{x} = \sum_{i=1}^{n} c_i\mathbf{a}_i, \quad c_i \in \mathbb{F}.$$

Proof: We will prove the linearly independent relation by contradiction, that is, suppose there exists $c, c_1, \ldots, c_n$ not all zero such that $\sum_{i=1}^{n} c_i\mathbf{a}_i + c\mathbf{b} = \mathbf{0}$. Since A is a linearly independent sequence, we must have that $c \neq 0$,

$$\sum_{i=1}^{n} c_i\mathbf{a}_i + c\mathbf{b} = \mathbf{0} \quad \Rightarrow \mathbf{b} = -\sum_{i=1}^{n} \frac{c_i}{c}\mathbf{a}_i$$

This is impossible since $\mathbf{b} \notin \text{Span}(A)$.

To show the second claim, suppose

$$\mathbf{x} = \sum_{i=1}^{n} c_i\mathbf{a}_i = \sum_{i=1}^{n} k_i\mathbf{a}_i, \quad c_i, k_i \in \mathbb{F}$$

Since A *is a linearly independent set*

$$\mathbf{0} = \sum_{i=1}^{n} (c_i - k_i)\mathbf{a}_i, \quad \Rightarrow \quad c_i - k_i = 0, \quad \Rightarrow \quad c_i = k_i$$

Hence, the expression is unique.

EXAMPLE 1.2.3

Let $V = M_{m\times n}(\mathbb{F})$. For any $1 \le i < m$ and $1 \le j \le n$, let $\mathbf{e}_{i,j}$ be the $m \times n$ matrix whose (i, j)-th entry is 1, and all other entries are zero. It is easy to see that $\mathbf{e}_{ij}$ for $1 \le i \le m$ and $1 \le j \le n$ are linearly independent since

$$\sum_{1\le i\le m, 1\le j\le n} c_i\mathbf{e}_{i,j} = \mathbf{0} \quad \Rightarrow \mathbf{c}_{i,j} = 0,$$

where $\mathbf{0}$ is the $m \times n$ matrix with entries all zero. Moreover, $M_{m\times n}(\mathbb{F})$ is spanned by the $\mathbf{e}_{ij}$ for $1 \le i \le m$ and $1 \le j \le n$.

1.3 Bases for Vector Spaces

Definition 1.3.1

A finite set subset $\mathcal{B} = \{\mathbf{v}_1, \ldots, \mathbf{v}_n\}$ **of a vector space** V is called a **finite basis** for V provided

1. $\mathbf{v}_1, \ldots, \mathbf{v}_n$ are linearly independent, and

2. $\mathbf{v}_1, \ldots, \mathbf{v}_n$ span V.

Consequently, if $\mathbf{v}_1, \ldots, \mathbf{v}_n$ is a list of vectors in V, then these vectors form a basis if and only if every $\mathbf{v} \in V$ can be uniquely written as

$$\mathbf{v} = a_1\mathbf{v}_1 + a_2\mathbf{v}_2 + \cdots + \mathbf{a}_n\mathbf{v}_n, \quad a_1, \ldots, a_n \in \mathbb{F}.$$

If there is an infinite set of linearly independent and spanning elements, then this set is called an **infinite basis**. This basis is called a Hamel basis.

EXAMPLE 1.3.1

We see in Example 1.2.3,

$$\mathcal{B} = \{\mathbf{e}_{i,j} \mid 1 \le i \le m,\ 1 \le j \le n\}$$

form a basis for the vector space $V = M_{m\times n}(\mathbb{F})$.

EXAMPLE 1.3.2

Let $V = \mathbb{F}[x]$. Then $\mathcal{B} = \{1 = x^0, x, x^2, \ldots\}$ form a basis of V.

Theorem 1.3.1

If V is generated by n vectors $\mathbf{v}_1,\ldots,\mathbf{v}_n$, then any sequence of vectors with more than n vectors is linearly dependent.

Proof: Let $(\mathbf{a}_1, \ldots, \mathbf{a}_k)$ with $k > n$ be a sequence of vectors of V. If $\mathbf{a}_1 \in V$, then $\mathbf{a}_1, \mathbf{v}_1,\ldots,\mathbf{v}_n$ are linearly dependent. There exists a vector say $\mathbf{v}_n \in V = \text{Span}(\mathbf{a}_1, \mathbf{v}_1,\ldots,\mathbf{v}_{n-1})$. We repeat this process, and obtain $V = \text{Span}(\mathbf{a}_n, \mathbf{a}_{n-1},\ldots,\mathbf{a}_1)$. Then $\mathbf{a}_{n+1} \in V$, *and* $\mathbf{a}_{n+1}, \mathbf{a}_n, \mathbf{a}_{n-1}, \ldots, \mathbf{a}_1$ are linearly dependent. Thus, we conclude that any sequence of vectors with more than n vectors is linearly dependent.

Theorem 1.3.2

Let V = $\text{Span}(\mathbf{v}_1,\ldots,\mathbf{v}_n)$ be a finitely generated vector space. Then V has a basis with at most n elements.

Proof: Let $\mathbf{v}_1,\ldots,\mathbf{v}_n$ generate V, that is $V = \text{Span}(\mathbf{v}_1,\ldots,\mathbf{v}_n)$. If $\mathbf{v}_1,\ldots,\mathbf{v}_n$ are linearly independent, then $\mathcal{B} = \{\mathbf{v}_1,\ldots,\mathbf{v}_n\}$ is a basis for V. If $\mathbf{v}_1,\ldots,\mathbf{v}_n$ are linearly dependent, then there exists a vector say $\mathbf{v}_n$ such that $\mathbf{v}_n \in \text{Span}(\mathbf{v}_1,\ldots, \mathbf{v}_{n-1})$. Then by Theorem 1.2.5, $V = \text{Span}(\mathbf{v}_1,\ldots, \mathbf{v}_{n-1})$. We repeat the process to obtain a set $\mathcal{B} = \{\mathbf{v}_1,\ldots,\mathbf{v}_m\}$ with $m < n$ such that $\mathcal{B}$ is a basis of V. Therefore, if $V = \text{Span}(\mathbf{v}_1,\ldots,\mathbf{v}_n)$, then the basis $\mathcal{B} = \{\mathbf{v}_1,\ldots,\mathbf{v}_m\}$ of V is such that $m \le n$.

From this proof, we see that one can obtain a basis from a spanning set. It is also easy to see that if V is a finitely generated vector space such that every linearly independent sequence from V has at most n vectors. Then V has a basis with at most n vectors. Moreover, if a vector space V has a basis with n elements, then any subspace of V has a basis with at most n elements.

Theorem 1.3.3

If $\mathcal{B}_1$ and $\mathcal{B}_2$ are two bases of a finitely generated vector space V, then $|\mathcal{B}_1| - |\mathcal{B}_2|$, i.e., the two bases contain the same number of elements.

Proof: Let $\mathcal{B}_1 - \{\mathbf{a}_1, \ldots, \mathbf{a}_m\}$ and $\mathcal{B}_2 - \{\mathbf{b1}, \ldots, \mathbf{b}_n\}$. Then $V - \text{Span}(\mathcal{B}_1)$, hence V has a basis with at most m elements, thus $n \le m$. On the other hand, $V - \text{Span}(\mathcal{B}_2)$, hence V has a basis with at most n elements, thus $m \le n$. Therefore, we must have that $m = n$. Hence, $|\mathcal{B}_1| - |\mathcal{B}_2|$, i.e., the two bases contain the same number of elements.

Suppose V is not finitely generated, i.e., has no finite basis. Let $\mathcal{B}_1$ and $\mathcal{B}_2$ be two bases with infinitely many elements. Consider the function $f : \mathcal{B}_1 \to \text{Span}(\mathcal{B}_2)$. For any $\mathbf{w} \in \mathcal{B}_1$, there exists a unique finite subset $\mathcal{B}_\mathbf{w} \subset \mathcal{B}_2$ such that $f(\mathbf{w}) = \sum_{I=1}^{N} c_I \mathbf{v}_I$ where $\{\mathbf{v}_1, \ldots, \mathbf{v}_n\} - \mathcal{B}_\mathbf{w}$. By cardinal arithmetic, $|\mathcal{B}_1| \ge |\bigcup_{w \in \mathcal{B}_1}, \mathcal{B}_\mathbf{w}| = |\mathcal{B}'|$ where $\mathcal{B}' - \bigcup_{\mathbf{w} \in B_1} \mathcal{B}_\mathbf{w}$. Since $\bigcup_{\mathbf{w} \in B_1} \mathcal{B}_\mathbf{w} \subseteq \mathcal{B}_2$, and both span V, we must have that $_{\mathbf{w}\in}$ $\mathcal{B}_\mathbf{w}$ $\mathcal{B}$ Therefore, $|\mathcal{B}_1| \ge |\mathcal{B}_2|$. Exchange the two bases, and with the same proof, we can show that $|\mathcal{B}_2| > |\mathcal{B}_1|$. Hence, when V has no finite bases, we still have $|\mathcal{B}_1| - |\mathcal{B}_2|$. We can conclude that the cardinality of any bases (finite or infinite) for a vector space over $\mathbb{F}$ are the same.

Definition 1.3.2

If V is a finitely generated vector space, then n, the number of vectors in a basis of V is called the **dimension** of V. We write $\dim V = n$. If V is not finitely generated, we denote $\dim V = \infty$.

It is easy to check the following results:

Theorem 1.3.4

Let V be a vector space and $\dim V - n$, and W is a subspace of V. Let $S = (\mathbf{v}_1, \ldots, \mathbf{v}_m)$ be a sequence of vectors from V. Then

1. If S is linearly independent, then $m \le n$.
2. If S spans V, then $m \ge n$.
3. If S is linearly independent and $m < n$, then S can be extended to a basis.
4. If S spans V and $m > n$, then some subsequence of S can form a basis.

5. If $m = n$ and S is linearly independent, then S spans V and S is a basis.

6. If $m = n$ and S spans V, then S is linearly independent and S is a basis.

7. $\dim(W) \leq n$; and $\dim W = n$ if and only if $W = V$.

8. There exists a subspace U of V such that $W + U = V$ and $W \cap U = \{\mathbf{0}\}$.

9. If W, U are subspaces of V, then $\dim(W) + \dim(U) = \dim(W + U) + \dim(W \cap U)$.

Definition 1.3.3

If V is a finite-dimensional vector space, an ordered basis for V is a finite sequence of vectors which is linearly independent and spans V. *If* $\mathcal{B} = \{\mathbf{b}_1, \ldots, \mathbf{b}_n\}$ is an ordered basis for V. Given $\mathbf{v} \in V$, then there is a unique n-tuple $(c_1, \ldots, c_n)$ such that $\mathbf{v} = \sum_{i=1}^{n} c_i \mathbf{b}_i$. Then c_i is called the *i-th* **coordinate of v relative to the ordered basis** $\mathcal{B}$. The **coordinate vector of v relative to** the ordered basis $\mathcal{B}$ is $\mathbf{c} = \begin{bmatrix} c_1 \\ \vdots \\ c_n \end{bmatrix}$. To indicate the dependence of this vector c on the basis, we denote this coordinate vector as $[\mathbf{v}]_{\mathcal{B}}$.

From this definition, one can proof the following results.

Theorem 1.3.5

Let V be an n-dimensional vector space over the field $\mathbb{F}$, and let $\mathcal{B} = \{\mathbf{b}_1, \ldots, \mathbf{b}_n\}$ and $\mathcal{B}' = \{\mathbf{b}_1', \ldots, \mathbf{b}_n'\}$ be two ordered bases of V, where $\mathbf{b}_i = [b_{i1}, \ldots, b_{in}]^T$ and $\mathbf{b}_i' = [b_{i1}' \ldots, b_{in}']^T$. Then there is a unique, necessarily invertible, $n \times n$ matrix P with entries in $\mathbb{F}$ such that

$$[\mathbf{v}]_{\mathcal{B}} = P[\mathbf{v}]_{\mathcal{B}'}; \quad [\mathbf{v}]_{\mathcal{B}'} = P^{-1}[\mathbf{v}]_{\mathcal{B}}, \quad \forall \mathbf{v} \in V.$$

The columns of P are given by

$$P_j = \left[\mathbf{b}_j'\right]_{\mathcal{B}}, \quad j = 1, \ldots, n. \quad \text{In fact } \mathrm{P} = \mathcal{B}^{-1}\mathcal{B}',$$

where $\mathcal{B}^{-1}$ means the inverse of the matrix formed by the basis elements.

Proof: We observe

$$\mathcal{B}[\mathbf{v}]_{\mathcal{B}} = \mathbf{v} = \mathcal{B}'[\mathbf{v}]_{\mathcal{B}'}$$

$$[\mathbf{v}]_{\mathcal{B}} = \mathcal{B}^{-1}\mathcal{B}'[\mathbf{v}]_{\mathcal{B}'}$$

$$= P[\mathbf{v}]_{\mathcal{B}'} \quad \text{where} \quad P = \mathcal{B}^{-1}\mathcal{B}' \text{ invertible matrix}$$

$$[\mathbf{v}]_{\mathcal{B}'} = P^{-1}[\mathbf{v}]_{\mathcal{B}}$$

$$P = \mathcal{B}^{-1}\mathcal{B}' \Rightarrow P_j = \mathcal{B}^{-1}\mathcal{B}'_j \Rightarrow \mathbf{b}'_j = \mathcal{B}\cdot P_j \Rightarrow P_j = \left[\mathbf{b}'_j\right]_{\mathcal{B}}.$$

Theorem 1.3.6

Suppose P is an $n \times n$ invertible matrix over $\mathbb{F}$. Let V be an n-dimensional vector space over $\mathbb{F}$, and let $\mathcal{B}$ be an ordered basis of V. Then there exits a unique ordered basis $\mathcal{B}'$ of V such that

$$[\mathbf{v}]_{\mathcal{B}} = P[\mathbf{v}]_{\mathcal{B}'}; \quad [\mathbf{v}]_{\mathcal{B}'} = P^{-1}[\mathbf{v}]_{\mathcal{B}}, \quad \forall\, \mathbf{v} \in V.$$

Proof: Let $\mathcal{B}' = P\mathcal{B}$, this is a basis since P and $\mathcal{B}$ are invertible matrices. By Theorem 1.3.10, $\mathcal{B}'$ satisfies the two stated conditions.

Suppose there exist two bases $\mathcal{B}'_1$ and $\mathcal{B}'_2$ and both satisfy the two stated conditions, then

$$[\mathbf{v}]_{\mathcal{B}} = P[\mathbf{v}]_{\mathcal{B}'_1} = P[\mathbf{v}]_{\mathcal{B}'_2} \quad \Rightarrow \quad [\mathbf{v}]_{\mathcal{B}'_1} = [\mathbf{v}]_{\mathcal{B}'_2}, \forall \mathbf{v}$$

Again by Theorem 1.3.5

$$I = \mathcal{B}_1'^{-1}\mathcal{B}'_2 \quad \Rightarrow \quad \mathcal{B}'_1 = \mathcal{B}'_2.$$

Thus such basis is unique.

EXAMPLE 1.3.3

Any finite set of vectors can be represented by a matrix in which its columns are the coordinates of the given vectors. As an example in dimension 2, consider a pair of vectors obtained by rotating the standard basis counterclockwise for 45°. The matrix whose columns are the coordinates of these vectors is

$$M = \begin{pmatrix} 1/\sqrt{2} & -1/\sqrt{2} \\ 1/\sqrt{2} & 1/\sqrt{2} \end{pmatrix}$$

1.4 Linear System of Equations

A linear equation is given as

$$a_1x_1 + a_2x_2 + \cdots + a_nx_n = b,$$

where x_1, x_2, … , x_n are variables, a_1, a_2, … , $a_n \in \mathbb{F}$ are coefficients of the equation and $b \in \mathbb{F}$ is the constant term. The coefficients of the equation can also be written in the form of the coefficient vector $\mathbf{a} = (a_1, a_2, \ldots, a_n)$. *The linear equation is called a homogeneous linear equation if the constant term is zero.*

A linear system of equations consists of m linear equations

$$\begin{cases} a_{11}x_1 + a_{12}x_2 + & \cdots & +a_{1n}x_n = b_1 \\ a_{21}x_1 + a_{22}x_2 + & \cdots & +a_{2n}x_n = b_2 \\ & \vdots & \\ a_{m1}x_1 + a_{m2}x_2 + & \cdots & +a_{mn}x_n = b_m \end{cases} \tag{1.1}$$

where x_1, x_2, … , x_n are variables of the linear system. For an x_i, we assume there at least exists a k such that $a_{ki} \neq 0$, otherwise the variable x_i can be eliminated from the linear system. A set of number s_1,…,s_n is a solution of the linear system (1.1), if all the linear equations become identical equations by substituting s_1, … ,s_n for x_i, …, x_n. The linear system is called a homogeneous linear system if all the constant terms are zero.

EXAMPLE 1.4.1

Consider three simple linear systems over a field $\mathbb{R}$

$$\begin{cases} x_1 + x_2 = 3 \\ x_1 - x_2 = 1 \end{cases}, \quad \begin{cases} x_1 + x_2 = 3 \\ 2x_1 + 2x_2 = 4 \end{cases}, \quad \text{and} \quad \begin{cases} x_1 + x_2 = 4 \\ x_1 + x_2 = 1 \end{cases}$$

The first linear system has the unique solution $x_1 = 2$, $x_2 = 1$, the second linear system has the solution $x_1 = c$, $x_2 = 2 - c$ where $c \in \mathbb{R}$ can be any constant, this solution has a free variable c and means infinite number of numerical solutions. The third linear system has no solution.

To solve a linear system of equations, a well known method is to transform the linear system to an equivalent simple linear system, a trapezoidal linear system based on Gaussian elimination. We consider three elementary transformations:

(i) Exchange two equations in the linear system;

(ii) Multiple by a non-zero constant to an equation in the linear system;

(iii) Add an equation to another equation in the linear system.

Theorem 1.4.1

For a linear system of equations, the solution set is invariant under any elementary transformations on the original linear system.

Proof: The elementary transformation (i) only changes the order of the individual equations, but the system is the same. Hence, the solution set stays the same under such transformation.

The elementary transformation (ii) only multiplies one individual equation by a non-zero scalar, producing an equivalent equation. Hence, the new system is equivalent to the original system, and the solution set is invariant under such transformation.

To show that the solution set is the same under elementary transformation (iii), we first let $c_1, \ldots, c_n$ *be a solution of the original linear system. Suppose we add the i-th equation to the j-th equation, i.e., the j-the equation is updated as*

$$(a_{i1} + a_{j1})x_1 + \ldots + (a_{in} + a_{jn})x_n = b_i + b_j$$

We need only show that $c_1, \ldots, c_n$ is the solution of this new equation. In fact from $a_{i1}c_1 + \ldots + a_{in}c_n - b_i$ and $a_{j1}c_1 + \ldots + a_{jn}c_n = b_j$ *we have*

$$(a_{i1} + a_{j1})_{c1} + \ldots + (a_{in} + a_{jn})c_n = b_i + b_j$$

Conversely, we start with the updated linear system. Update the *i-th equation by multiplying by* -1, using elementary transformation (II), we get a new linear system having the same solution. Update this new linear system by adding the j-th equation to the updated i-th equation, we get back the original linear system having the same solution similar to the above discussion.

Notice that $a_{i1} \neq 0$ for some i and one can exchange the first equation and the i-the equation if $a_{11} = 0$. Therefore, we may assume $a_{11} \neq 0$ in linear system (1.1), we multiple $-a_{j1}/a_{11}$ to the first equation and add the result to the j-th equation for $j = 2, \ldots, n$. After the transformation, we obtain a new linear system (to avoid many symbols we here still use the a_{ij} and b_j to represent the updated coefficients).

$$\begin{cases} a_{11}x_1 + a_{12}x_2 + \cdots + a_{1n}x_n = b_1 \\ \qquad a_{2k}x_k + \cdots + a_{2n}x_n = b_2 \\ \qquad \vdots \\ \qquad a_{mk}x_k + \cdots + a_{mn}x_n = b_m \end{cases}$$

where k is the least number associated to x_k whose coefficient a_{2k} is non-zero in the updated $m - 1$ equations and $k \geq 2$. Note that if there is no such k, then the system is equivalent to the first equation. Otherwise, repeat the similar transformation process for the updated $m - 1$ equations, we obtain

$$\begin{cases} a_{11}x_1 + a_{12}x_2 + a_{13}x_3 + \cdots + a_{1n}x_n = b_1 \\ a_{2k}x_k + a_{2k+1}x_{k+1} + \cdots + a_{2n}x_n = b_2 \\ a_{3l}x_1 + \cdots + a_{3n}x_n = b_3 \\ \vdots \\ a_{ml}x_l + \cdots + a_{mn}x_n = b_m \end{cases}$$

where $l \geq k + 1$ and $a_{3l} \neq 0$. Repeat the transformation process and we finally can get the linear system in the trapezoidal form

$$\begin{cases} a_{11}x_1 + a_{12}x_2 + a_{13}x_3 + \cdots + a_{1n}x_n = b_1 \\ a_{2k}x_k + a_{2k+1}x_{k+1} + \cdots + a_{2n}x_n = b_2 \\ a_{3l}x_1 + \cdots + a_{3n}x_n = b_3 \\ \vdots \\ a_{rs}x_s + \cdots + a_{rn}x_n = b_r \\ 0 = b_{r+1} \\ \vdots \\ 0 = b_m \end{cases} \tag{1.2}$$

where $a_{11}, a_{2k}, a_{3l}, \ldots, a_{rs}$ are all non-zero and $1 < k < l < \ldots < s \leq n$, $r \leq m$. Notice that there is no equation of the form $0 = b_i$ if $r = m$. For this trapezoidal linear system, variables $x_{s+i}, \ldots, x_n$ are free or independent variables, and the variables $x_i, x_k, x_l, \ldots, x_s$ are the dependent variables with nonzero coefficients, which can be expressed in terms of the free variables.

Theorem 1.4.2

The linear system of equations (1.2) has solutions if and only if $b_{r+1i} = \cdots = b_m = 0$. Furthermore, the solution is unique if and only if $r = n$ and $b_{r+i} = \ldots = b_m = 0$; there are an infinite number of solutions if and only if $r < n$ and $b_{r+1} = \cdots = b_m = 0$.

Proof: Suppose $b_{r+1i} = \ldots = b_m = 0$. Fix the free variables with random values $c_r +_i, \ldots, c_n$ and substitute them into the *r-th* equation, we have $a_{rs}x_x + b'_r = b_r$ where b'_r is a number in $\mathbb{F}$. Then $x_s = c_s = \dfrac{b_r - b'_r}{a_{rs}}$ is solved easy. Similarly, substitute $c_r, c_{r+1}, \ldots, c_n$ into the $r - 1$-th equation we solve $x_l - c_{r-1}$ for some variable x_l. Repeat the similar process for i-th equation $i = r - 2, \ldots, 1$ we will find a solution $c_1, \cdots, c_r, c_{r+1}, \ldots, c_n$ *of* (1.2).

If one of $b_{r+1}, \ldots, b_m$ *is non-zero w.l.o.g*, $b_m \neq 0$, then the equation $0 = b_m$ is a contradictory equation and leads to no solution for (1.2).

If $r = n$ and $b_{r+1} = \ldots = b_m = 0$, then there is no free variables and the above solving process shows that the solution is unique.

If $r < n$ and $b_{r+1} = \ldots = b_m = 0$, then then free variables can be any value, i.e., there are infinite number of solutions.

Two linear systems are equivalent if they have the same solutions or no solution simultaneously. By Theorem 1.4.1 and Theorem 1.4.2, we have

Corollary 1.4.1. A linear system is equivalent to a trapezoid linear system.

1.5 A First Look at Determinants

Consider a simple linear system

$$\begin{cases} a_{11}x_1 + a_{12}x_2 = b_1 \\ a_{21}x_1 + a_{22}x_2 = b_2 \end{cases} \quad a_{ij} \neq 0 \tag{1.3}$$

Multiply the first equation by a_{22}, and multiply the second equation by a_{12}, we have

$$a_{11}a_{22}x_1 + a_{12}a_{22}x_2 = b_1a_{22},$$
$$a_{12}a_{21}x_1 + a_{12}a_{22}x_2 = b_2a_{12}.$$

Subtracting the second equation from the first equation, we get a new equation with x_2 being eliminated

$$(a_{11}a_{22} - a_{12}a_{21})x_1 = b_1a_{22} - b_2a_{12}.$$

Similarly, x_i can be eliminated in the system (1.3) to get

$$(a_{11}a_{22} - a_{12}a_{21})x_2 = b_2a_{11} - b_1a_{21}.$$

Then we can solve for x_1 *and* x_2 if $a_{11}a_{22} - a_{12}a_{21} \neq 0$. Precisely,

$$x_1 = \frac{b_1a_{22} - b_2a_{12}}{a_{11}a_{22} - a_{12}a_{21}} \quad \text{and} \quad \frac{b_2a_{11} - b_1a_{21}}{a_{11}a_{22} - a_{12}a_{21}}$$

For numbers a, b, c, d, we can consider that $ad - bc$ is a value computed from the matrix $\begin{pmatrix} a & b \\ c & d \end{pmatrix}$. We call this value the *determinant* of the matrix and denote it by $\begin{vmatrix} a & b \\ c & d \end{vmatrix}$.

Proposition 1.5.1. If $a_{11}a_{22} \neq a_{12}a_{21} \neq 0$, then the solution of (1.3) *is*

$$x_1 = \frac{\begin{vmatrix} b_1 & a_{12} \\ b_2 & a_{22} \end{vmatrix}}{\begin{vmatrix} b_{11} & a_{12} \\ b_{21} & a_{22} \end{vmatrix}}, \quad x_2 = \frac{\begin{vmatrix} a_{11} & b_1 \\ a_{21} & b_2 \end{vmatrix}}{\begin{vmatrix} a_{11} & a_{12} \\ a_{21} & a_{22} \end{vmatrix}}$$

The matrix consisting of the coefficients of a linear system and its determinant play import roles in solving the linear system, more details can be found in following chapters.

1.6 Using Computer Algebra Systems to Perform Computations

Computer software programs such as Mathematica [Wol15], Maple [Map16], Sage [SJ05], and R [R C13] incorporate algorithms to simplify and compute linear systems. In this section, we include some of the computational examples provided by students using a variety of software programs.

For example, to perform a row reduction on a rectangular matrix, "Row Reduce" function gives the row reduced matrix of any matrix.

```
In[1]:= RowReduce[{{1, 2, 3, 1, 0, 0}, {4, 5, 6, 0, 1, 0},
        {7, 8, 9, 0, 0,1}}]
Out[1]= {{1, 0, -1, 0, -8/3, 5/3}, {0, 1, 2, 0, 7/3, -4/3},
        {0, 0, 0, 1, -2, 1}};
```

One of the most important uses of matrices is to represent and solve linear systems. This section will discuss how to solve linear systems with Mathematica [Wol15].

Solving a linear system involves solving a matrix equation $Ax = \mathbf{b}$, where $A = \{a_{ij}\}$ is an $m \times n$ matrix and $\mathbf{b} = \{b_i\}$ is a column vector given in (1.1). This is a set of m linear equations in n unknowns. If $m = n$, then the system is said to be square. If $m > n$, i.e., there are more equations than unknowns, then the system is said to be over determined. If $m < n$, i.e., there are fewer equations than unknowns, then the system is said to be under determined.

The "LinearSolve" function in Mathematica [Wol15] is designed to solve linear systems directly.

Find the solution to $\mathbf{Ax} = \mathbf{b}$ where A is a square matrix with exact arithmetic:

```
In[1]:= A = {{1, 1, 1}, {1, 2, 3}, {1, 4, 9}};
b = {1, 2, 3};
LinearSolve[A, b]
Out[1]={-1/2, 2, -1/2}
```

Find a solution for a rectangular matrix:

```
In[1]:= A={{1, 2, 3, 4}, {5, 6, 7, 8}, {9, 10, 11, 12}};
b={13, 14, 15};
LinearSolve[A, b]
Out[1]= {-25/2,51/4,0,0}
```

The "Linear Solve" function is not limited to solve a system when $\mathbf{b}$ is a column vector, it also solves $\mathbf{Ax} = \mathrm{b}$ for x when $\mathbf{b}$ is a matrix:

```
In[1]:= A = {{1, 1, 1}, {1, 2, 3}, {1, 4, 9}};
b = {{1, 2}, {3, 4}, {5, 6}};
LinearSolve[A, b]
Out[1]={{-2,-1},{4,4},{-1,-1}}
```

Sage [SJ05] is a free computer algebra system. Following is an example to obtain an echelon form of a given matrix.

```
M = random_matrix(ZZ, 3, 4)
print M
[-2 -1  1  2]
[ 6  1  3  0]
[-2  2  1 -7]
M.echelon_form()
[ 2  0 11  1]
[ 0  1  6 -3]
[ 0  0 18  0]
```

Three elementary row operations to a given matrix can also be obtained via Sage. The following example shows how to multiply a non-zero constant to an equation in the linear system, exchange two equations in the linear system, or add an equation to another equation in the linear system.

```
C=matrix(QQ, [[2,3,5,7],[9,3,4,6],[7,9,2,1],[3,0,4,8]]); C
[2 3 5 7]
[9 3 4 6]
[7 9 2 1]
[3 0 4 8]
D=C.with_rescaled_row(0,3); D
[ 6  9  15 21]
[ 9  3   4  6]
[ 7  9   2  1]
[ 3  0   4  8]
E=C.with_swapped_rows(0,2); E
[7 9 2 1]
[9 3 4 6]
[2 3 5 7]
[3 0 4 8]
F=C.with_added_multiple_of_row(1,3,2); F
[ 2  3  5  7]
[15  3 12 22]
[ 7  9  2  1]
[ 3  0  4  8]
```

Another potential use of Sage is looking at the basis of vector spaces along with looking at the basis of the intersection of vector spaces.

```
V = (QQ^3).span([vector(QQ,[1,2,3]),vector(QQ,[3,4,9])])
W = (QQ^3).span([vector(QQ,[4,6,12]),vector(QQ,[3,2,12])])
print("Basis of V")
V.basis_matrix()
print("Basis of W")
W.basis_matrix()
print("Basis of the intersection")
V.intersection(W).basis_matrix()

Basis of V
[1 0 3]
[0 1 0]
Basis of W
[ 1 0 24/5]
[ 0 1 -6/5]
Basis of the intersection
[ 1 3/2 3]
```

R [R C13] is a programming language and free software environment for statistical computing and graphics that is supported by the R Foundation for Statistical Computing. We may use R in linear algebra. For example, we may compute the determinate of a matrix by R.

```
[language=R]
M=matrix(c(7,4,6,3,8,4,7,15,2,6,3,9,-3,6,-17,3),nrow=4,ncol=4)
M
     [,1]  [,2] [,3]   [,4]
[1,]    7     8    2     -3
[2,]    4     4    6      6
[3,]    6     7    3    -17
[4,]    3    15    9      3
det(M)
[1] -6930
```

Using *R*, one can also test linear independence relationships among given vectors, and determine the rank of the vector spaces.

```
[language=R]
N = matrix(c(1,0,0,0,2,5,4,-1,-1,0,1,2,0,-10,7,3,-4,-10,-8,2),
nrow=4,ncol=5,dimnames=list(c("A","B","C","D"),
paste("v",1:5,sep="")))
N
  v1 v2 v3  v4  v5
A  1  2 -1   0  -4
B  0  5  0 -10 -10
C  0  4  1   7  -8
D  0 -1  2   3   2
mN <- N[,q$pivot[seq(q$rank)]]
mN
  v1 v2 v3  v4
A  1  2 -1   0
B  0  5  0 -10
C  0  4  1   7
D  0 -1  2   3
qr(N)$rank
[1] 4
```

1.7 Applications of Vector Spaces

Vector spaces are applied throughout mathematics, science, and engineering. Vector spaces may be generalized in several ways, leading to more advanced notions in geometry and abstract algebra. In this section, we will discuss a few applications.

1.7.1 Code

In this section, we will see how vectors can be used to design codes for detecting errors that may occur in data transmission. A *linear code* of length n and rank k is a linear subspace C with dimension k of the vector space $\mathbb{F}_q^n$ where $\mathbb{F}_q$ is the finite field with q elements. Such a code is called a q-ary *code*. If $q = 2$ or $q = 3$, the code is described as a *binary code* or a *ternary code*, respectively. The vectors in C are called *codewords*. The size of a code is the number of codewords and equals q^k.

In practice, to transmit a message consisting of words, numbers, or symbols, one begins by encoding each "word" of the message as a binary vector, that is, a vector with entries in $\mathbb{Z}_2$. A *binary code* is a set of binary vectors of the same length called *code vectors*. The process of converting a message into code vectors is called *encoding*, and the reverse process is called *decoding*.

For example, a message "buy" is represented by a binary code 1011. One way of encoding is to attach a binary "tail" to the binary code to detect the error—attaching a "1" if the binary code contains an odd number of 1s, otherwise, attaching a "0". Thus, all encoded binary words have an even number of 1's. Therefore, 1011 will be encoded as 10111, and we know that an error has occurred if the word is distorted to 00111. Therefore, the message to be transmitted consists of binary vectors and a simple error-detecting code called a *parity check code*. A parity check code is created by appending an extra component, called a *check digit*, to each vector so that the parity, i.e., the total number of 1's is even.

To generalize this process in terms of vectors, let the message be the binary vector $\mathbf{a} = [a_1, \ldots, a_n] \in \mathbb{Z}_2^n$, and the parity check code vector is $\mathbf{v} = [a_1, \ldots, a_n, a] \in \mathbb{Z}_2^{n+1}$, where the check digit a is chosen such that

$$a_1 + \ldots + a_n + a \equiv 0 \in \mathbb{Z}_2, \text{ equivalently } \mathbf{1} \cdot \mathbf{v} \equiv 0, \mathbf{1} = [1, \ldots, 1] \in \mathbb{Z}_2^{n+1}.$$

The vector $\mathbf{1} = [1, 1, \ldots, 1]$, is called a *check vector*. Parity check codes are special cases of the more general *check digit codes* in a q-ary code.

For example, let a message be represented by a vector $\mathbf{a} = [a_1, \ldots, a_n] = [2, 2, 0, 1, 2] \in \mathbb{Z}_3^5$, and the parity check code vector is $\mathbf{v} = [2, 2, 0, 1, 2, a] \in \mathbb{Z}_3^6$, then the check digit must be

$$\mathbf{1} \cdot \mathbf{v} = 2 + 2 + 1 + 2 + a \equiv 1 + a \equiv 0 \in \mathbb{Z}_3 \quad \Rightarrow \quad a = 2.$$

Thus the parity check code becomes $\mathbf{v} = [2, 2, 0, 1, 2, 2]$.

Simple check digit code may detect a single error, but it will not be able to detect other types of common errors such as the accidental permuting of two digits. For example, the binary codes 1011 and 0111 both contain an odd number of 1's so the added parity check code cannot detect the error if the message "buy" represented by a binary code 1011 is distorted to 0111. Next, we will discuss how to use matrices to design codes that can correct as well as detect certain types of errors.

First, we encode a message represented by a vector $\mathbf{a} \in \mathbb{Z}_2^k$ by using a matrix transformation $T : \mathbb{Z}_2^k \to \mathbb{Z}_2^n$, for some $n > k$. The vector $T(\mathbf{a}) = G\mathbf{a} = \mathbf{v}$ is called the *code vector,* and the matrix G is called the *generator matrix* for the code. To exam whether a received vector $\mathbf{b}$ is a code vector, we define a *parity check matrix* P to be such that $P\mathbf{v} = 0$ for the code.

For example, a message "sell" represented by a binary code vector $\mathbf{a} = [0101]^T$ is encode by the generator matrix G to get the code vector $\mathbf{v}$ as the following

$$T : \mathbb{Z}_2^4 \to \mathbb{Z}_2^7, \quad T(\mathbf{a}) = G\mathbf{a} = \begin{bmatrix} 1 & 0 & 0 & 0 \\ 0 & 1 & 0 & 0 \\ 0 & 0 & 1 & 0 \\ 0 & 0 & 0 & 1 \\ 1 & 1 & 0 & 1 \\ 1 & 0 & 1 & 1 \\ 0 & 1 & 1 & 1 \end{bmatrix} \begin{bmatrix} 0 \\ 1 \\ 0 \\ 1 \end{bmatrix} = \begin{bmatrix} 0 \\ 1 \\ 0 \\ 1 \\ 0 \\ 1 \\ 0 \end{bmatrix} = \mathbf{v}$$

The parity check matrix

$$P = \begin{bmatrix} 1 & 1 & 0 & 1 & 1 & 0 & 0 \\ 1 & 0 & 1 & 1 & 0 & 1 & 0 \\ 0 & 1 & 1 & 1 & 0 & 0 & 1 \end{bmatrix}$$

If the vector $\mathbf{v}$ is received then $P\mathbf{v} = 0$, and it is correct. On the other hand, if $\mathbf{v}' = [0111010]^T$ is received, then compute

$$P\mathbf{v}' = \begin{bmatrix} 1 & 1 & 0 & 1 & 1 & 0 & 0 \\ 1 & 0 & 1 & 1 & 0 & 1 & 0 \\ 0 & 1 & 1 & 1 & 0 & 0 & 1 \end{bmatrix} \begin{bmatrix} 0 \\ 1 \\ 1 \\ 1 \\ 0 \\ 1 \\ 0 \end{bmatrix} = \begin{bmatrix} 0 \\ 1 \\ 1 \end{bmatrix}$$

Since $P\mathbf{v}' \neq \mathbf{0}$, the code is distorted. Furthermore, $P\mathbf{v}' = [0\ 1\ 1]^T$ is the same as the third column of the parity check matrix P, and this tells us the error is in the third component of the received message $\mathbf{v}'$. Thus, by changing the third component of $\mathbf{v}'$, we recover the correct code vector $\mathbf{v}$.

1.7.2 Hamming Code

In telecommunication, *Hamming codes* are a family of linear error-correcting codes that were invented by Richard Hamming in 1950. The binary Hamming code first introduced has check matrix

$$H = \begin{bmatrix} 0 & 0 & 0 & 1 & 1 & 1 & 1 \\ 0 & 1 & 1 & 0 & 0 & 1 & 1 \\ 1 & 0 & 1 & 0 & 1 & 0 & 1 \end{bmatrix}$$

The elements of this matrix are in $\mathbb{Z}_2$. The Hamming code C of length 7 is the space of H, i.e.,

$$C = \{\mathbf{v} \in \mathbb{Z}_2^7 \mid H\mathbf{v} = \mathbf{0}.\}$$

The *Hamming distance* d_H between any two words of the same length is defined as the number of coordinates in which they differ.

Proposition 1.7.1. The minimum distance between binary Hamming codewords is 3.

Proof: Suppose $\mathbf{x}, \mathbf{y} \in C$ are two code words from a Hamming code. Then $\mathbf{x} - \mathbf{y} \in C$.

If $d_H(\mathbf{x}, \mathbf{y}) = \mathbf{1}$, then $H(\mathbf{x} - \mathbf{y})$ is a column of H, which is non-zero. But this is not possible since if $(\mathbf{x} - \mathbf{y})$ is a Hamming code word, then $H(\mathbf{x} - \mathbf{y}) = 0$.

If $d_H(\mathbf{x}, \mathbf{y}) = 2$, then $H(\mathbf{x} - \mathbf{y}) = \mathbf{0}$ if and only if there are two columns of H which are linearly dependent. But this is not the case.

Hence $d_H(\mathbf{x}, \mathbf{y}) \geq 3$ for all code words $\mathbf{x}, \mathbf{y}$. Every check matrix for a binary Hamming code will have three columns that are linearly dependent, so in fact some code words are of distance 3.

Linear codes like C are identified by their length, dimension, and minimal distance. Thus C is referred as (7, 4, 3), because its length is 7, its dimension is 4, and its minimum distance is 3.

1.7.3 Linear Code

Below, we give formal definition and theorem to generalize this idea.

Definition 1.7.1.

If $k < n$, then any $n \times k$ matrix of the form $G = \begin{bmatrix} \mathbb{I}_k \\ A \end{bmatrix}$ where A is an $(n-k) \times k$ matrix over $\mathbb{Z}_2$ is called a **standard generator matrix** for an $[n, k]$ **binary code** $T : \mathbb{Z}_2^k \to \mathbb{Z}_2^n$. Any $(n-k) \times n$ matrix of the form $P = [B\ \mathbb{I}_{n-k}]$ where B is an $(n-k) \times k$ matrix over $\mathbb{Z}_2$ is called a **standard parity check matrix**. The code is said to have length n and dimension k.

The following theorem will help us to identify when G can be a standard generator matrix for an *error-correcting* binary code, and also, given G, tell us how to find an associated standard parity check matrix P.

Theorem 1.7.1

If $G = \begin{bmatrix} \mathbb{I}_k \\ A \end{bmatrix}$ is a standard generator matrix and $P = [B\ \mathbb{I}_{n-k}]$ is a standard parity check matrix, then P is the parity check matrix associate with G if and only if $A = B$. The corresponding binary code $[n, k]$ is (single) error-correcting if and only if the columns of P are non-zeros and distinct.

Proof: Let $\mathbf{a}_i$ be the *i-th column of A*.

($\Rightarrow$) Let P be the standard parity check matrix associated with G for the same binary code **a.** Then for every $\mathbf{x} \in Z_2^k$, we have $PG\mathbf{x} = \mathbf{0}$. Thus,

$$PG\mathbf{x} = [B\mathbb{I}_{n-k}]\begin{bmatrix} \mathbb{I}_k \\ A \end{bmatrix}\mathbf{x} = (B\mathbb{I} + \mathbb{I}A)\mathbf{x} = B\mathbf{x} + A\mathbf{x} = \mathbf{0}, \quad \forall \mathbf{x} \in \mathbb{Z}_2^k$$

Hence

$$B\mathbf{x} = -A\mathbf{x} = A\mathbf{x} \quad \textbf{over } \mathbb{Z}_2.$$

Let $\mathbf{x} = \mathbf{e}_i$, the i-th standard basis vector in $\mathbb{Z}_2^k$, then we have

$$\mathbf{b}_i = B\mathbf{e}_i = A\mathbf{e}_i = \mathbf{a}_i, \quad \forall i, \Rightarrow B = A.$$

($\Leftarrow$) If $B = A$, then

$$PG\mathbf{x} = [B\ \mathbb{I}_{n-k}]\begin{bmatrix} \mathbb{I}_k \\ A \end{bmatrix}\mathbf{x} = (B + A)\mathbf{x} = 2A\mathbf{x} = 0, \quad \forall \mathbf{x} \in \mathbb{Z}_2^k$$

Thus, P is the standard parity check matrix associated with G for the same binary code **a.**

Finally, to see that the pair G, P determines an error-correcting code if the columns of P are non-zeros and distinct, let $\mathbf{x}$ be a message vector in $\mathbb{Z}_2^k$, and let the corresponding code vector be $\mathbf{v} = G\mathbf{x}$, and $P\mathbf{v} = \mathbf{0}$.

(⇐) Suppose there is an error in the i-th component of the received code vector $\mathbf{v}'$. Then $\mathbf{v}' = \mathbf{v} + \mathbf{e}_i$, and

$$P\mathbf{v}' = P(\mathbf{v} + \mathbf{e}_i) = P\mathbf{v} + P\mathbf{e}_i = \mathbf{0} + \mathbf{p}_i = \mathbf{p}_i$$

where $\mathbf{p}_i$ is the i-th column of P, which indicates exact component that the error occurs.

(⇒) If P has a zero column, say $\mathbf{p}_i = \mathbf{0}$, then an error in the i-th column will not be detected, since $P\mathrm{v}' = 0$. Moreover, if $\mathbf{p}_i = \mathbf{p}_j$, then we cannot pinpoint the column that the error occurs.

We observer that $G = \begin{bmatrix} \mathbb{I}_k \\ A \end{bmatrix}$ and $P = [B\ \mathbb{I}_{n-k}]$ guarantee that the columns of G and the rows of P are linearly independent. Theorem 1.7.3 states that G and P are associated with the same code if and only if $A = B$, which is equivalent to the condition that $PG = \mathbb{O}$. Moreover, $PG = \mathbb{O}$ means that $Pg_i = \mathbf{0}$ for every column g_i of the matrix G. Thus, vector $\mathbf{v}$ is a code vector if and only if $\mathbf{v}$ is in the column space of the matrix G, that is $\mathbf{v} = G\mathbf{a}$ for some vector $\mathbf{a} \in \mathbb{Z}_2^k$. Since elementary row or column operations do not affect the row or column spaces of the given matrix. Therefore, if P is a parity check matrix, E is an elementary matrix, and $\mathbf{v}$ a code vector, then

$$EP\mathbf{v} = E(P\mathbf{v}) = E\mathbf{0} = \mathbf{0},$$

and EP is also a parity check matrix. Therefore, any parity check matrix can be converted into another one by means of a sequence of row operations.

The properties of row or column operation of a matrix in linear algebra provide ways of construction new code. Below, we give a new definition:

Definition 1.7.2

For $n > k$, and an $n \times k$ matrix G and an $(n-k) \times n$ matrix P over $\mathbb{Z}_2^k$ are **generator matrix** and a **parity check matrix,** for an $[n, k]$ binary code C if the following conditions are all satisfied:

1. The columns of G are linearly independent.
2. The rows of P are linearly independent.
3. $PG = \mathbb{O}$.

We call two codes C_1 and C_2 equivalent if there is a permutation matrix M such that

$$\{\mathbf{Mx} : \mathbf{x} \in C_1\} = C_2.$$

1.7.4 Golay Code

Golay codes are the most interesting codes constructed up to 1996. In mathematics and electronics engineering, a binary Golay code is a type of linear error-correcting code used in digital communications. Golay codes were used to transmit the color images of Jupiter and Saturn within a constrained telecommunications bandwidth by the Voyager 1 and 2 spacecraft.

Definition 1.7.3

(*Extended binary Golay codes*) Let G be the matrix $G = [\mathbb{I}_{12} A]^T$, where $\mathbb{I}_{12}$ is the 12×12 identity matrix and A is the 12×12 matrix

$$A = \begin{bmatrix} 0 & 1 & 1 & 1 & 1 & 1 & 1 & 1 & 1 & 1 & 1 & 1 \\ 1 & 1 & 1 & 0 & 1 & 1 & 1 & 0 & 0 & 0 & 1 & 0 \\ 1 & 1 & 0 & 1 & 1 & 1 & 0 & 0 & 0 & 1 & 0 & 1 \\ 1 & 0 & 1 & 1 & 1 & 0 & 0 & 0 & 1 & 0 & 1 & 1 \\ 1 & 1 & 1 & 1 & 0 & 0 & 0 & 1 & 0 & 1 & 1 & 0 \\ 1 & 1 & 1 & 0 & 0 & 0 & 1 & 0 & 1 & 1 & 0 & 1 \\ 1 & 1 & 0 & 0 & 0 & 1 & 0 & 1 & 1 & 0 & 1 & 1 \\ 1 & 0 & 0 & 0 & 1 & 0 & 1 & 1 & 0 & 1 & 1 & 1 \\ 1 & 0 & 0 & 1 & 0 & 1 & 1 & 0 & 1 & 1 & 1 & 0 \\ 1 & 0 & 1 & 0 & 1 & 1 & 0 & 1 & 1 & 1 & 0 & 0 \\ 1 & 1 & 0 & 1 & 1 & 0 & 1 & 1 & 1 & 0 & 0 & 0 \\ 1 & 0 & 1 & 1 & 0 & 1 & 1 & 1 & 0 & 0 & 0 & 1 \end{bmatrix}, A^T = A$$

The binary linear code with generator matrix G is called the **extended binary Golay code** and will be denoted by G_{24}.

Proposition 1.7.1 (Properties of the extended binary Golay code)

1. The length of G_{24} is 24 and its dimension is 12.

2. A parity-check matrix for G_{24} is the 12×24 matrix $H = [A\mathbb{I}_{12}]$.

3. The code G_{24} is self-dual, i.e., $G_{24}^{\perp} = G_{24}$.

4. Another parity-check matrix for G_{24} is $[\mathbb{I}_{12}\ \mathrm{A}] = G$.

5. Another generator matrix for G_{24} is matrix $[\mathrm{A}\ \mathbb{I}_{12}]^T$.

6. The minimum distance of G_{24} is 8.

Proof: We will leave the proof as an exercise for the readers. Observed that a generator matrix for the dual code is a parity-check matrix for the original code and vice versa. The dual of the dual code is always the original code. For more details concerning item 3, please see the following discussion concerning dual code.

In coding theory, the dual code of a linear code C is the linear code defined by

$$C^1 = \left\{ x \in \mathbb{Z}_2^n \mid \langle \mathbf{x}, \mathbf{c} \rangle = \sum_{i=1}^{n} x_i c_i = 0, \quad \forall \mathbf{c} \in C \right\}$$

Proposition 1.7.2 Let C be an $[n, k]$ binary code with generator matrix G Then $C^{\perp}$ is an $[n, n-k]$ code with parity check matrix G^T.

Proof: $\mathbf{x} \in C^{\perp} \quad \Leftrightarrow \quad \mathbf{x} \cdot \mathbf{c} = 0, \quad \forall \mathbf{c} \in C \quad \Leftrightarrow \quad \mathbf{x} \cdot (G\mathbf{a}) = 0, \quad$ for all $\mathbf{a} \in \mathbb{Z}_2^k$

But

$$\mathbf{x} \cdot (G\mathbf{a}) = (G\mathbf{a})^T \mathbf{x} = \mathbf{a}^T G^T \mathbf{x} = a^T (G^T \mathbf{x}) = 0, \forall\ \mathbf{a} \in \mathbb{Z}_2^k \quad \Leftrightarrow \quad G^T \mathbf{x} = \mathbf{0},$$

This means that $C^{\wedge}$ is the null space of G^T, and G^T is an $k \times n$ matrix of rank k. Hence $C^{\perp}$ has dimension $n - k$. G^T is a parity check matrix of $C^{\perp}$.

As a conclusion, vector spaces are very important in coding theory. The theory of error-correcting codes is a relatively recent application of mathematics to information and communication systems. It turns out that a rich set of mathematical ideas and tools from linear and abstract algebra can be used to design good codes.

1.8 Exercises

1. Identify which of the following sets are subspaces of $\mathbb{R}^3$, and state the reasons.

(*a*) $V = \{(x, y, z) \in \mathbb{R}^3 \mid x + 2y = 11\}$.

(*b*) $V = \{(x, y, z) \in \mathbb{R}^3 \mid x + 2y + 5z = 0\}$.

(*c*) $V = \{(r, r + 2,0) \in \mathbb{R}^3\}$.

(*d*) The set of all polynomials $f(x)$, with $f(7) = 0$.

2. Assume that **v** and w are linear independent vectors. Prove that **v, w,** (**v** + **w**) are linear dependent vectors.

3. Prove that all polynomials $f(x)$ with degree not greater than 3 constitute a vector space.

4. Given $V = \{(2, 5, 3), (1, 0, 2)\}$ and $W = \{(2, 0, 5), (3, 5, 5)\}$. Is the intersection of Span(V) and Span(W) a vector space? If so, what is the dimension?

5. Let $B = \{(1, 0, 1), (0, 2, 0), (1, 2, 3)\}$ and $C = \{(1, 0, 0), (2, 0,1), (0,1, 3)\}$ be two bases of $\mathbb{R}^3$. Let the coordinates of a vector **v** relative to B and C be (x, y, z) and (a, b, c). Write the relation between these coordinates in matrix notation.

6. Let U and V both be 2-dimensional subspaces of $\mathbb{R}^5$, and let $W = U \cap V$. Find all possible values for the dimension of W.

7. Let U and V both be 2-dimensional subspaces of $\mathbb{R}^5$, and define $W = U + V = \{w \mid w = u + v, u \in U, v \in V\}$. Show that W is a linear space, and find all possible values for the dimension of W.

8. Let $A : \mathbb{R}^n \to \mathbb{R}^k$ be a real matrix, not necessarily square. Show that if the columns of A are linearly dependent, then A is not one-to-one.

9. Find a 3×3 matrix that acts on $\mathbb{R}^3$ as follows: it keeps the x-axis fixed but rotates the yz-plane by 60 degrees.

10. Let $A = \begin{bmatrix} -1 & -1 \\ 3 & 3 \end{bmatrix}$. Find all 2×2 matrices, B such that $AB = 0$. Does $\{B \in M_{2\times2} \mid AB = 0\}$ form a subspace of 2×2 matrices? If so, find a basis.

11. Let V be the vector space over all real 2×2 matrices. Let W be the subset of V consisting of all symmetric matrices. (a) Prove that W is a subspace of V. (b) Find a basis of W. (c) Determine the dimension of W.

12. A matrix A is called skew-symmetric if $A^T = A$. Let V be the vector space over all real 2×2 matrices. Let W be the subset of V consisting of all skew-symmetric matrices. (a) Prove that W is a subspace of V. (b) Find a basis of W. (c) Determine the dimension of W.

13. A Vandermonde matrix is often used in applications. Calculate the square of the Vandermonde matrix

$$\begin{bmatrix} 1 & 1 & 1 & \cdots & 1 \\ x_1 & x_2 & x_3 & \cdots & x_n \\ x_1^2 & x_2^2 & x_3^2 & \cdots & x_n^2 \\ \vdots & \vdots & \vdots & \vdots & \vdots \\ x_1^{n-1} & x_2^{n-1} & x_3^{n-1} & \cdots & x_n^{n-1} \end{bmatrix}$$

14. Calculate the images resp. the pre-images of the line

$$ax + by + c = 0$$

and the conic section

$$ax^2 + 2bxy + cy^2 = 0$$

under the affine mapping

$$(x, y) \to (a_{11}x + a_{12}y + C_1, a_{21}x + a_{22}y + c_2)$$

where $a_{11}a_{22} - a_{12}a_{21} = 0$.

15. Show that if $f: \mathbb{R}^2 \to \mathbb{R}^2$ is an injective, affine mapping, then f maps conic sections onto conic sections. For which f is the image of a circle also a circle?

16. If Q is the conic

$$\{(x, y) \in \mathbb{R}^2 \mid a_{11}x^2 + 2a_{12}xy + a_{22}y^2 + 2b_1x + 2_{b2}y + c = 0\}$$

and

$$A = \begin{bmatrix} a_{11} & a_{12} \\ a_{12} & a_{22} \end{bmatrix}, \quad D = \begin{bmatrix} a_{11} & a_{12} & b_1 \\ a_{21} & a_{22} & b_2 \\ b_1 & b_2 & c \end{bmatrix}$$

Show that the conic has the form

$$\{(x, y) : X^T DX = 0\}, \text{ where } X = \begin{bmatrix} x \\ y \\ 1 \end{bmatrix}$$

17. Show that if z_1 and z_2 are distinct complex numbers, then

$$\det \begin{bmatrix} z & \overline{z} & 1 \\ z_1 & \overline{z}_1 & 1 \\ z_2 & \overline{z}_2 & 1 \end{bmatrix} = 0$$

is the equation of the straight line through z_1 and z_2

18. For which values of t do the following 4-vectors form a basis of $\mathbb{R}^4$?

$$(1, t, 3, 4), (t, t, 3, 4t), (1, t, 3t, 4), (1, 1, 3, 4t).$$

19. For which values of t the following matrices form a basis in $M_{2\times2}$

$$\begin{bmatrix} t & 2t \\ 2 & 3t \end{bmatrix}, \begin{bmatrix} 1 & 2 \\ 2t & 3 \end{bmatrix}, \begin{bmatrix} 1 & 2t \\ t+1 & t+2 \end{bmatrix}, \begin{bmatrix} 1 & t+1 \\ 2 & 2t+1 \end{bmatrix}.$$

CHAPTER 2

LINEAR TRANSFORMATIONS

2.1 Linear Transformations

In mathematics, a linear map, or a linear transformation, or, in some contexts, a linear function, is a mapping V a W between two algebraic structures, that preserves the operations of addition and scalar multiplication. Linear maps can often be represented as matrices, and simple examples include rotation and reflection linear transformations. One of the important special cases is that when $V = W$, the map is called a linear operator, or an endomorphism of V. In the language of abstract algebra, a linear map is a module homomorphism of modules over a given ring. In fact, ring and module are regarded as one of the most important subjects in abstract algebra and the theory of ring and module is so extensive which includes that vector spaces. In this chapter, we will focus on linear transformations of vector spaces, and touch a few topics in ring and module homomorphisms.

Recall, if a function f maps from a set X, namely the domain into a set Y, codomain, then for each $x \in X$, the element $f(x) \in Y$ is called the image of x, and $\{x \in X \mid f(x) = y\}$ is the pre-image of y. The range of f, denoted by range(f), is the set of all images, range(f) $= \{f(x) \mid x \in X\}$. A linear function, or linear transformation, or a linear map between vector spaces is a function that preserves some of the algebraic properties of vector spaces.

Definition 2.1.1

Let V and W be vector spaces over the field F. A linear transformation from V into W is a function T from V into W such that

$$T(c\mathbf{v} + \mathbf{w}) = cT(\mathbf{v}) + \mathbf{T(w)},\ \forall \mathbf{v}, \mathbf{w} \in V, c \in \mathbb{F}.$$

The set of all linear transformation from V to W will be denoted by $\mathcal{L}(V, W)$.

EXAMPLE 2.1.1

Consider $L : \mathbb{R}^2 \to \mathbb{R}^3$, a linear transformation, defined by $L([x, y]^T) = [y, -2x + 2y, x]^T$, then this linear transformation can be given as a matrix multiplication such that

$$L([x,y]^T) = A\begin{bmatrix} x \\ y \end{bmatrix} \quad \text{where } A = \begin{bmatrix} 0 & 1 \\ -2 & 2 \\ 1 & 0 \end{bmatrix}$$

In algebra, a module homomorphism is a function between modules that preserves module structure.

Definition 2.1.2

Let M and N be left modules over a ring R, then a function $f : M \to N$ is called a **module homomorphism** or a R-**linear map** if for any $x, y \in M$ and $r \in R$,

1. $f(x + y) = f(x) + f(y)$,
2. $f(rx) = rf(x)$.

If M, N are right modules, then the second condition is replaced with $f(xr) = f(x)r$. The pre-image of the zero element under f is called the kernel of f or the **null space of** f. The set of all module homomorphisms from $M \to N$ is denoted by $\mathrm{Hom}_R(M, N)$. It is an abelian group with respect to addition, but is not necessarily a module unless R is commutative.

Remark: In general, if $T \in \mathrm{Hom}\,(M, N)$, then

1. $\ker(T) = \{m \in M \mid T(m) = 0\}$.
2. $\mathrm{im}(T) = \{n \in N \mid \exists\, m \in M, T(m) = n\}$.
3. T is injective, or monomorphism, or one-to-one, if $\ker(T) = \{\mathbf{0}\}$.
4. T is surjective, or epimorphism, or onto, if $\mathrm{im}(T) = N$.
5. T is bijective, or isomorphism, if T is both injective and surjective.
6. M and N are said to be isomorphic, $M \simeq N$ if there exists an isomorphism $T \in \mathrm{Hom}(M, N)$.

Since a vector space is a module over a field, therefore, linear transformations of vector spaces are special cases of module homomorphisms.

EXAMPLE 2.1.2

Let $R = \mathbb{Z}$ be a ring. Let $M = \mathbb{Z}_m$, and $N = \mathbb{Z}_n$ be two R-modules. Let $f : M \rightarrow N$ be a R-module homomorphism. One can check that $\mathrm{Hom}_R(M, N) \cong \mathbb{Z}_{\gcd(m;\, n)}$.

Definition 2.1.3

If V is a vector space over the field F, a **linear operator** on V is a linear transformation from V into V.

Definition 2.1.4

A linear transformation T is **non-singular** if $T(\gamma) = \mathbf{0}$ implies that $\gamma = \mathbf{0}$. That is, the null space of T is $\{\mathbf{0}\}$.

Theorem 2.1.1

Let T be a linear transformation from V into W. Then T is non-singular if and only if T carries each linearly independent subset of V onto a linearly independent subset of W.

Proof: Let $T : V \rightarrow W$ be a linear transformation and non-singular, and $\mathbf{v}_1, \ldots, \mathbf{v}_n$ linearly independent. To show $T(\mathbf{v}_1), \ldots, T(\mathbf{v}_n)$ are linearly independent, consider

$$\begin{aligned} c_1T(\mathbf{v}_1) + \ldots + c_nT(\mathbf{v}_n) &= \mathbf{0} && \text{where } c_i \in \mathbb{F}, \\ T(c_1\mathbf{v}_1 + \ldots + c_n\mathbf{v}_n) &= \mathbf{0} && \text{since } T \text{ is a linear transformation} \\ c_1\mathbf{v}_1 + \ldots + c_n\mathbf{v}_n &= && \mathbf{0} \text{ since } T \text{ is non-singular} \\ c_1 = \ldots = c_n &= \mathbf{0} && \text{since } \mathbf{v}_1, \ldots, \mathbf{v}_n \text{ are linearly independent.} \end{aligned}$$

Thus, $T(\mathbf{v}_1), \ldots, T(\mathbf{v}_n)$ are linearly independent.

To show T is non-singular, we only need to show that $T(\mathrm{v}) = \mathbf{0}$ **implies** $\mathbf{v} = \mathbf{0}$. To do so, let $\mathbf{v}_1, \ldots, \mathbf{v}_n$ be a basis of V, and $\mathbf{v} = c_1\mathbf{v}_1 + \ldots + c_n\mathbf{v}_n$, then

$$T(\mathbf{v}) = T(c_1\mathbf{v}_1 + \ldots + c_n,\mathbf{v}_n) = \mathbf{0}$$

$$\mathbf{0} = c_1T(\mathbf{v}_1) + \ldots + c_nT(\mathbf{v}_n) \Rightarrow c_1 = \ldots = c_n = 0$$

since $T(\mathbf{v}_1), \ldots, T(\mathbf{v}_n)$ are linearly independent

$$\mathbf{v} = c_1\mathbf{v}_1 + \ldots + c_n\mathbf{v}_n = \mathbf{0}$$

Thus T is non-singular.

EXAMPLE 2.1.3

In 2-dimensional space $\mathbb{R}^2$ linear maps are described by 2×2 real matrices relative to the standard basis. These are some examples:

1. Rotation by 90 degrees counterclockwise:
$$A = \begin{pmatrix} 0 & -1 \\ 1 & 0 \end{pmatrix}$$
2. Rotation by angle θ counterclockwise:
$$A = \begin{pmatrix} \cos\theta & -\sin\theta \\ \sin\theta & \cos\theta \end{pmatrix}$$
3. Reflection against the x-axis:
$$A = \begin{pmatrix} 1 & 0 \\ 0 & -1 \end{pmatrix}$$
4. Reflection against the y-axis:
$$A = \begin{pmatrix} -1 & 0 \\ 0 & 1 \end{pmatrix}$$
5. Scaling by $k > 0$ in all directions:
$$A = \begin{pmatrix} k & 0 \\ 0 & k \end{pmatrix}$$
6. Horizontal shear mapping:
$$A = \begin{pmatrix} 1 & m \\ 0 & 1 \end{pmatrix}$$
7. Vertical squeeze mapping, $k > 1$:
$$A = \begin{pmatrix} 1 & 0 \\ 0 & 1/k \end{pmatrix}$$
8. Projection onto the y-axis:
$$A = \begin{pmatrix} 0 & 0 \\ 0 & 1 \end{pmatrix}$$

2.2 Rank and Nullity of a Linear Transformation

Definition 2.2.1

Let V and W be vector spaces over the field $\mathbb{F}$, and let T be a linear transformation from V into W. The **null space** of T is the set of all vectors

$\mathbf{v} \in V$ such that $T(\mathbf{v}) = \mathbf{0}$. If V is finite-dimensional, the **rank** of T is the dimension of the range of T, and the **nullity** of T is the dimension of the null space of T.

EXAMPLE 2.2.1

Let $\mathbb{R}[s]$ denote the ring of polynomials in a single variable s with coefficients from the field $\mathbb{R}$. A polynomial vector $\mathbf{v} \in (\mathbb{R}[s])^w$ is a vector of size w with each entry being a polynomial. The degree n of a vector $\mathbf{v} \in (\mathbb{R}[s])^w$ is the maximum amongst the degrees of its polynomial components. Alternatively we can write $\mathbf{v}$ as a polynomial of degree n with the coefficients being the vectors from $\mathbb{R}^w$. Hence, $(\mathbb{R}[s])^w = \mathbb{R}^w[s]$. Similarly a polynomial matrix $\mathbb{R} \in \mathbb{R}^{g\times w}[s]$ is a matrix of size $g \times w$ with the entries from $\mathbb{R}[s]$. The degree of a polynomial matrix is the maximum of the degrees amongst its polynomial entries. A polynomial matrix can be written as a polynomial in s with coefficients being the matrices from $\mathbb{R}^{g\times w}$. The null space of $\mathbb{R}$ is $\{\mathbf{v} \in \mathbb{R}^w[s] \mid R\mathbf{v} = \mathbf{0}\}$.

Theorem 2.2.1

Let V and W be finite-dimensional vector spaces over the field $\mathbb{F}$, let $T \in \mathcal{L}(V, W)$ be a linear transformation from V into W. Then

1. If T is injective, then $\dim V \leq \dim W$.
2. If T is surjective, then $\dim V \geq \dim W$.
3. If $\dim V = \dim W$, then T is an isomorphism if and only if T is injective or T is surjective.
4. $\text{rank}T + \text{nullity}T = \dim V$.

Proof: Let $\mathcal{B} = \{\mathbf{b}_1, \ldots, \mathbf{b}_n\}$ be a basis for V.

1. Since T is injective, $T(\mathbf{b}_1), \ldots, T(\mathbf{b}_n)$ are a linearly vectors in W. Hence independent $n \leq \dim W$.
2. Since T is surjective, $\{T(\mathbf{b}_1), \ldots, T(\mathbf{b}_n)\}$ generates W, i.e., $W = \text{Span}(T(\mathbf{b}_1), \ldots, T(\mathbf{b}_n))$. Hence $n \geq \dim(W)$.
3. ($\Rightarrow$) If T is an isomorphism, then it is both injective and bijective. ($\Leftarrow$) If T is injective, then $T(\mathbf{b}_1), \ldots, T(\mathbf{b}_n)$ are linearly independent. Since $\dim V = \dim W$, $T(\mathbf{b}_1), \ldots, \mathrm{T}(\mathbf{b}_n)$ form a basis for W. Hence T is surjective. Therefore, T is an isomorphism.

On the other hand, if T is surjective, then $W = \text{Span}(T(\mathbf{b}_1), \ldots, T(\mathbf{b}_n))$, so $\{T(\mathbf{b}_1), \ldots, T(\mathbf{b}_n)\}$ contains a basis. Since $\dim V = \dim W$, $T(\mathbf{b}_1), \ldots, T(\mathbf{b}_n)$ in fact form a basis. Let $T(\mathbf{v}) = \mathbf{0}$ **for some** $\mathbf{v} = \sum_{i=1}^{n} c_i \mathbf{b}_i$. Then $\mathbf{0} = T(\mathbf{v}) = T\left(\sum_{i=1}^{n} c_i \mathbf{b}_i\right) = \sum_{i=1}^{n} c_i T(\mathbf{b}_i)$ implies all $c_i = 0$. Thus $\mathbf{0} = \sum_{i=1}^{n} c_i \mathbf{b}_i = \mathbf{0}$. Thus $\ker T = \{\mathbf{0}\}$, hence T is injective. Therefore, T is an isomorphism.

4. Let $k = \text{nullity}T$, and let $\mathbf{b}_1, \ldots, \mathbf{b}_k$ be the basis for $\ker T$, and let $\mathbf{b}_1, \ldots, \mathbf{b}_k, \mathbf{b}_{k+1}, \ldots, \mathbf{b}_n$ is a basis for V. We will show that rank $T = n - k$, that means we need to show that $T(\mathbf{b}_{k+1}), \ldots, T(\mathbf{b}_n)$ form a basis for $\text{im}T$. To do so, we let $w \in \text{im}T$, so there exists $\mathbf{v} \in V$ such that $T(\mathbf{v}) = \mathbf{w}$. Thus,

$$\mathbf{w} = T(\mathbf{v}) = T\left(\sum_{i=1}^{n} c_i \mathbf{b}_i\right) = T\left(\sum_{i=1}^{k} c_i \mathbf{b}_i\right) + \sum_{i=k+1}^{n} c_i T(\mathbf{b}_i) = \sum_{i=k+1}^{n} c_i T(\mathbf{b}_i),$$

where the last equality holds since $\mathbf{b}_i$ for $i = 1, \ldots, k$ are the basis for $\ker T$. Thus $\text{im}T \subseteq \text{Span}(T(\mathbf{b}_{k+1}), \ldots, T(\mathbf{b}_n))$. Moreover, $T(\mathbf{b}_{k+1}), \ldots, T(\mathbf{b}_n)$ must be linearly independent, otherwise, there would be c_i for $i = k + 1, \ldots, n$, not all zero, such that $\sum_{i=k+1}^{n} c_i T(\mathbf{b}_i) = T\left(\sum_{i=k+1}^{n} c_i \mathbf{b}_i\right) = \mathbf{0}$, which implies that $\sum_{i=k+1}^{n} c_i \mathbf{b}_i \in \ker(T)$. But this is impossible, since $\mathbf{b}1, \ldots, \mathbf{b}_k$ form a basis for $\ker T$, which in turn would imply $\mathbf{b}_1, \ldots, \mathbf{b}_n$ are linearly dependent, contradicting the fact they form a basis for V. Therefore, $T(\mathbf{b}_{k+1}), \ldots, T(\mathbf{b}_n)$ are linearly independent and span $\text{im}T$, hence $\text{rank}T = \dim(\text{im}T) = n - k$. Thus, $\dim V = n = k + (n - k) = \text{nullity}T + \text{rank}T$.

Theorem 2.2.2

(Sylvester's theorem) Let V, W and Y be finitely generated vector space over a field $\mathbb{F}$, and $a \in \mathcal{L}(V, W)$, $\beta \in \mathcal{L}(W, Y)$ are linear transformations. Then

1. $\text{nullity}(\beta\alpha) \leq \text{nullity}(\alpha) + \text{nullity}(\beta)$.

2. $\text{rank}(\alpha) + \text{rank}(\beta) - \dim W \leq \text{rank}(\beta\alpha) \leq \min\{\text{rank}(\alpha), \text{rank}(\beta)\}$.

Proof: To prove the first claim, we let $\beta' = \beta|_{\text{im}(\alpha)}$ the restriction of to $\text{im}(\alpha)$. Then $\ker(\beta') \subseteq \ker(\beta)$, therefore nullity $(\beta') \leq \text{nullity}(\beta)$. Hence,

$$\begin{aligned}\text{nullity}(\beta\alpha) &= \dim(V) - \mathit{rank}(\beta\alpha)\\ &= [\dim(V) - \text{rank}(\alpha)] + [\text{rank}(\alpha) - \text{rank}(\beta\alpha)]\\ &= \text{nullity}(\alpha) + \text{nullity}(\beta') \le \text{nullity}(\alpha) + \text{nullity}(\beta).\end{aligned}$$

To show the second claim, we note that $\text{im}(\beta\alpha) \subseteq \text{im}(\beta)$, so

$$\text{rank}(\beta\alpha) = \dim(\text{im}(\beta\alpha)) < \dim(\text{im}(\beta)) = \text{rank}(\beta).$$

Furthermore, since $\text{im}(\beta\alpha) = \text{im}(\beta')$, and rank of any linear transformation is not greater than the dimension of its domain, we must have $\text{rank}(\beta\alpha) \le \text{rank}(\alpha)$. Thus,

$$\text{rank}(\beta\alpha) \le \min\{\text{rank}(\alpha), \text{rank}(\beta)\}.$$

$$\begin{aligned}&\text{rank}(\alpha) + \text{rank}(\beta) - \dim(W)\\ &= [\dim(V) - \text{nullity}(\alpha)] + [\dim(W) - \text{nullity}(\beta)] - \dim(W)\\ &= \dim(V) - [\text{nullity}(\alpha) + \text{nullity}(\beta)] \qquad \text{by claim 1}\\ &\le \dim(V) - \text{nullity}(\beta\alpha)\\ &= \text{rank}(\beta\alpha).\end{aligned}$$

Hence, we proved the claim.

Remark

Let $P \in \mathcal{L}(V, W)$, $A \in \mathcal{L}(W, Y)$, and $Q \in \mathcal{L}(Y, Z)$, where V, W, Y, Z are finitely generated vector spaces. Then

1. If P is surjective, then $\text{rank}A = \text{rank}(AP)$.
2. If Q is injective, then $\text{rank}A = \text{rank}(QA)$.
3. If P, Q are isomorphisms, then $\text{rank}A = \text{rank}(QAP)$.

Definition 2.2.2

The **column rank** of an $m \times n$ matrix A is the dimension of the subspace of $\mathbb{F}^m$ spanned by the columns of A. Similarly, the **row rank** is the dimension of the subspace of the space $\mathbb{F}^n$ of row vectors spanned by the rows of A.

Theorem 2.2.2

If A is an $m \times n$ matrix with entries in the field $\mathbb{F}$, then

$$\text{row rank }(A) = \text{column rank }(A)$$

Proof: To show this, we need to observe that there exist an invertible $m \times n$ matrix Q and an invertible $n \times n$ matrix P such that $A_1 = QAP$ has the block form $A_1 = \begin{bmatrix} I & 0 \\ 0 & 0 \end{bmatrix}$ where I is an $r \times r$ identity matrix for some r, and the rest of the matrix is zero. The matrices P, Q in fact are the products of elementary matrices, i.e., the row or column operations on A. (Note that P and Q are invertible matrices, hence they are isomorphisms, and preserve rank.) For this matrix, it is obvious that row rank = column rank = r. The strategy is to reduce an arbitrary matrix to this form.

Remark: Another proof for the claim rank$T$ + nullity$T$ = dim V:

Let A be an $m \times n$ matrix with r linearly independent columns, that is rank(A) = r. We will produce an $n \times (n - r)$ matrix X whose columns form a basis of the null space of A. Thus the nullity of A is equal to the rank of X.

Without loss of generality, assume that the first r columns of A are linearly independent. So, we can write $A = [A_1\ A_2]$, where A_1 is $m \times r$ with r linearly independent column vectors and A_2 is $m \times (n - r)$, each of whose $n - r$ columns are linear combinations of the columns of A_1, that is, $A_2 = A_1B$ for some $r \times (n - r)$ matrix B. Hence, $A = [A_1\ A_1\ B]$. Let

$$X = \begin{pmatrix} -B \\ I_{n-r} \end{pmatrix},$$

where I_{n-r} is the $(n - r) \times (n - r)$ identity matrix, so X is an $n \times (n - r)$ matrix such that

$$AX = [A_1 A_1 B]\begin{pmatrix} -B \\ I_{n-r} \end{pmatrix} = -A_1B + A_1B = 0$$

Therefore, each of the $n - r$ columns of X is a particular solution of $\mathbf{Ax} = \mathbf{0}$. Furthermore, the $n - r$ columns of X are linearly independent because $X\mathbf{u} = \mathbf{0}$ will imply $\mathbf{u} = \mathbf{0}$:

$$X\mathbf{u} = \mathbf{0} \Rightarrow \begin{pmatrix} -B \\ I_{n-r} \end{pmatrix}\mathbf{u} = \mathbf{0} \Rightarrow \begin{pmatrix} -B\mathbf{u} \\ \mathbf{u} \end{pmatrix} = \begin{pmatrix} \mathbf{0} \\ \mathbf{0} \end{pmatrix} \Rightarrow \mathbf{u} = \mathbf{0}$$

Therefore, the column vectors of X form a set of $n - r$ linearly independent solutions for $A\mathbf{x} = \mathbf{0}$.

Definition 2.2.3

Short exact sequences are exact sequences of the form

$$0 \to A \xrightarrow{f} B \xrightarrow{g} C \to 0$$

where f is an injective map and g is a surjective map, and the image of f is equal to the kernel of g.

The short exact sequence is called split if there exists a homomorphism $h : C \to B$ such that the composition $g \circ h$ is the identity map of C. It follows that $B \cong A \oplus C$.

If $0 \to A \xrightarrow{f} B \xrightarrow{g} C \to 0$ is a short exact sequence of vector spaces, then $\dim A + \dim C = \dim B$. In addition, $A \cong \ker(g)$ and $C = \mathrm{im}(g)$.

To generalize this, we have if

$$0 \to V_1 \to V_2 \to \cdots \to V_r \to 0$$

is an exact sequence of finite-dimensional vector spaces, then

$$\sum_{i=1}^{r} (-1)^i \dim(V_i) = 0$$

2.3 Representation of Linear Transformations by Matrices

Usually, a linear operator can be represented by a matrix. Let V be a finite dimensional vector space over the field $\mathbb{F}$, and let $\mathcal{B} = \{\mathbf{b}1, \ldots, \mathbf{b}_n\}$ be an ordered basis for V. This linear transformation can be represented by a matrix P, where the columns of P are given by

$$P_j = [T(\mathbf{b}_j)]_{\mathcal{B}}, \quad j = 1, \ldots, n.$$

This matrix P is called a **matrix representation** of the linear transformation T.

EXAMPLE 2.3.1

Let $T : \mathbb{R}^3 \to \mathbb{R}^3$ be defined by $T(a_1, a_2, a_3) = (3a_1 + a_2, a_1 + a_3, a_1 - a_3)$. Consider the standard ordered basis $\varepsilon = \{e_1, e_2, e_3\}$. With respect to this basis the coordinate vector of an element $\mathbf{a} = (a_1, a_2, a_3)$ is $[\mathbf{a}]_\varepsilon = [a_1, a_2, a_3]^T$.

The matrix representation of T is $\begin{pmatrix} 3 & 1 & 0 \\ 1 & 0 & 1 \\ 1 & 0 & -1 \end{pmatrix}$.

On the other hand, if we choose a basis $\mathcal{B} = \{\mathbf{b}_1 = (1, 0, 0), \mathbf{b}_2 = (1, 1, 0), \mathbf{b}_3 = (1, 1, 1)\}$. Then the coordinate vectors of $\mathbf{a}$ and $T(\mathbf{a})$ are $[\mathbf{a}]_{\mathcal{B}} = [a_1 - a_2, a_2 - a_3, a_3]^T$, and $[T(\mathbf{a})]_{\mathcal{B}} = [2a_1 + a_2 - a_3, 2a_3, a_1 - a_3]^T$. The columns of matrix P are the coordinate vectors of $T(\mathbf{b}_1)$, $T(\mathbf{b}_2)$, $T(\mathbf{b}_3)$ relative to the basis $\mathcal{B}$.

$$T(\mathbf{b}_1) = \begin{bmatrix} 3 \\ 1 \\ 1 \end{bmatrix}_{\mathcal{E}} = \begin{bmatrix} 2 \\ 0 \\ 1 \end{bmatrix}_{\mathcal{B}}, \quad T(\mathbf{b}_2) = \begin{bmatrix} 4 \\ 1 \\ 1 \end{bmatrix}_{\mathcal{E}} = \begin{bmatrix} 3 \\ 0 \\ 1 \end{bmatrix}_{\mathcal{B}}, \quad T(\mathbf{b}_3) = \begin{bmatrix} 4 \\ 2 \\ 0 \end{bmatrix}_{\mathcal{E}} = \begin{bmatrix} 2 \\ 2 \\ 0 \end{bmatrix}_{\mathcal{B}}$$

The matrix representation of T with respect to this basis is $\begin{pmatrix} 2 & 3 & 2 \\ 0 & 0 & 2 \\ 1 & 1 & 0 \end{pmatrix}$.

Definition 2.3.1

Let $\mathcal{B} = \{\alpha_1, \ldots, \alpha_n\}$ be an ordered basis for V, and $\beta' = \{\beta_1, \ldots, \beta_m\}$ be an ordered basis for W. If T is any linear transformation from V into W, then T is determined by its action on the vectors α_j:

$$T(\alpha_j) = \sum_{i=1}^{m} A_{ij}\beta_i, \quad [T]_{\mathcal{B}\to\mathcal{B}'} = \left[T(\alpha_j)\right]_{\mathcal{B}\to\mathcal{B}'} = (A_{1j}, \cdots, A_{mj})^T, j = 1, \cdots n.$$

Then the matrix $A = (A_{ij})$ of size $m \times n$ is called the matrix of T relative to the pair of ordered basis $\mathcal{B}$ and $\mathcal{B}'$.

Theorem 2.3.1

Let V and W be finite-dimensional vector spaces over $\mathbb{F}$ of dimension n and m respectively. Let $\mathcal{B}$ and $\mathcal{B}'$ be the basis for V and W. Then the map $\phi : \mathcal{L}(V, W) \to M_{m\times n}(\mathbb{F})$ defined by the matrix

$\phi(T) = [T]_{\mathcal{B}\to\mathcal{B}'} \in M_{m\times n}(\mathbb{F})$. Then the following diagram commutates

$$\begin{array}{ccc} V & \xrightarrow{T} & W \\ \downarrow \cong & & \downarrow \cong \\ \mathbb{F}^n & \xrightarrow{(\ )} & \mathbb{F}^m \end{array}$$

where the isomorphisms are the coordinate mappings.

In this setting, the following result follows directly.

Theorem 2.3.2

Let V be a finite-dimensional vector space over the field $\mathbb{F}$, and let

$$\mathcal{B} = \{a_1, \cdots, a_n\}, \mathcal{B}' = \{\alpha'_1, \cdots, \alpha'_n\}$$

be the ordered bases for V. Suppose T is a linear operator on V. If $P = [P_1, \ldots, P_n]$ is the $n \times n$ matrix with columns $P_j = [\alpha'_j]_{\mathcal{B}}$, then

$$[T]_{\mathcal{B}' \to \mathcal{B}'} = P^{-1}\,[T]_{\mathcal{B} \to \mathcal{B}}\,P$$

Alternatively, if U is the invertible operator on V defined by $U(\alpha_j) = \alpha'_j$, $j = 1, \ldots, n$, then

$$T_{\mathcal{B}' \to \mathcal{B}'} = [U]_{\mathcal{B}}^{-1}\,[T]_{\mathcal{B} \to \mathcal{B}}\,[U]_{\mathcal{B}}$$

Proof: The proof follows from the diagram below:

$$\begin{array}{ccc}
(V, \mathcal{B}) & \xrightarrow{[T]_{\mathcal{B} \to \mathcal{B}}} & (V, \mathcal{B}) \\
P \big\downarrow & & \big\downarrow P^{-1} \\
(V, \mathcal{B}') & \xrightarrow{[T]_{\mathcal{B}' \to \mathcal{B}'}} & (V, \mathcal{B}')
\end{array}$$

Definition 2.3.2

Let A, B be $n \times n$ matrices over the field $\mathbb{F}$. We say that B is similar to A over $\mathbb{F}$ if there is an invertible $n \times n$ matrix P over $\mathbb{F}$ such that $B = P^{-1}AP$.

Furthermore, from the above definition, we can see that if we let V be an n-dimensional vector space over the field $\mathbb{F}$ and W an m-dimensional *vector space over* $\mathbb{F}$. Let $\mathcal{B}$ be an ordered basis for V and $\mathcal{B}'$ be an ordered basis for W. For each linear transformation T from V into W, there is an $m \times n$ matrix A with entries in $\mathbb{F}$ such that

$$[T(\alpha)]_{\mathcal{B}'} = A[\alpha]_{\mathcal{B}}, \quad \forall \alpha \in V.$$

Moreover, for fixed bases $\mathcal{B}$ and $\mathcal{B}'$, $T \to A$ is a one-one correspondence between the set of all linear transformations from V into W and the set of all $m \times n$ matrices over the field $\mathbb{F}$.

This matrix A is called the **matrix of T relative to the ordered bases $\mathcal{B}$ and $\mathcal{B}'$**. In case $V = W$ *and* $\mathcal{B} = \mathcal{B}'$, we call A **matrix of T relative to the ordered basis $\mathcal{B}$**.

From the above definition, we can show that

Theorem 2.3.3

Let V be an m-dimensional vector space over the field $\mathbb{F}$ and W an m-dimensional vector space over $\mathbb{F}$. Let $\mathcal{B}$ be an ordered basis for V and $\mathcal{B}'$ be an ordered basis for W. For each pair of ordered basis $\mathcal{B}$ and $\mathcal{B}'$ for V, W respectively, the function which assigns to a linear transformation T its matrix relative to $\mathcal{B}$ and $\mathcal{B}'$ is a bijection between $\mathcal{L}(V, W)$ and the space of all $m \times n$ matrices over the field $\mathbb{F}$.

Proof: Let $\phi : \mathcal{L}(V, W) \to M_{m\times n}(\mathbb{F})$, where $\phi(T) = [T]_{\mathcal{B}\to\mathcal{B}'}$. We need to show this map is one-to-one and onto.

To show ϕ is one-to-one, note that, since every linear transformation is uniquely defined by its evaluation on the basis. If $\phi(T) = \phi(S)$ for some $T, S \in \mathcal{L}(V, W)$, then $[T]_{\mathcal{B}\to\mathcal{B}'} = [S]_{\mathcal{B}\to\mathcal{B}'}$, which means every column $[T(\mathbf{b}_i)]_{\mathcal{B}'} = [S(\mathbf{b}_i)]_{\mathcal{B}'}$ for $i = 1, \ldots, n$. Thus $T(\mathbf{b}_i) = S(\mathbf{b}_i)$ for every $\mathbf{b}_i \in \mathcal{B}$. Therefore, $T = S$.

To see ϕ is onto, let $A = \{a_{ji}\} \in M_{m\times n}(\mathbb{F})$. Define a linear transformation $T : V \to W$ by $T(\mathbf{b}_i) = \sum_{j=1}^{n} a_{ji}\mathbf{b}'_j$, hence $[T(\mathbf{b}_i)]_{\mathcal{B}'} = [a_{11}, \ldots, a_{mi}]^T$. Thus $\phi(T) = A$, therefore, ϕ is onto.

In the next section, we will show that $\mathcal{L}(V, W)$ is a vector space, and $\phi : \mathcal{L}(V, W) \to M_{m\times n}(\mathbb{F})$ is a linear transformation. Hence, Theorem 2.4.1 implies $\mathcal{L}(V, W)$ @ $M_{m\times n}(\mathbb{F})$, and the following results follow directly.

Theorem 2.3.4

Let V be an n-dimensional vector space over the field $\mathbb{F}$, and let W be an m-dimensional vector space over $\mathbb{F}$. Then the space $\mathcal{L}(V, W)$ is finite-dimensional and has dimension mn.

Theorem 2.3.5

Let V, W, Z be finite-dimensional vector spaces over the field $\mathbb{F}$. Let T be a linear transformation from V into W, and U be a linear transformation from W into Z. If $\mathcal{B}$, $\mathcal{B}'$ and $\mathcal{B}^2$ for V, W, Z respectively, if A is the matrix of T relative to the pair $\mathcal{B}$, $\mathcal{B}^2$, and B is the matrix of U relative to the pair $\mathcal{B}'$, $\mathcal{B}^2$, then the matrix of the composition UT relative to the pair $\mathcal{B}$, $\mathcal{B}^2$ is the product matrix $C = BA$.

2.4 The Algebra of Linear Transformation

Let V be a finite-dimensional vector space over the field $\mathbb{F}$, let T and U be a linear transformation from V into W. For any $\mathbf{v} \in V$, the function $(T + U)$ defined by

$$(T + U)(\mathbf{v}) = T(\mathbf{v}) + U(\mathbf{v})$$

is a linear transformation from V into W. *If* $c \in \mathbb{F}$, the function cT defined by

$$(cT)(\mathbf{v}) = c(T(\mathbf{v})),$$

is a linear transformation from V into W. The set of all linear transformation from V into W, together with the addition and scalar multiplication defined above, is a vector space over the field $\mathbb{F}$.

Theorem 2.4.1

For each pair of ordered basis $\mathcal{B}$ and $\mathcal{B}'$ for V, W respectively, the function which assigns to a linear transformation T its matrix relative to $\mathcal{B}$ and $\mathcal{B}'$ is a bijection between $\mathcal{L}(V, W)$ and the space of all $m \times n$ matrices over the field $\mathbb{F}$.

Lemma Let V be an n-dimensional vector space over the field $\mathbb{F}$ and W an m-dimensional vector space over $\mathbb{F}$. Let $\mathcal{B}$ be an ordered basis for V and $\mathcal{B}'$ be an ordered basis for W. Let $\phi : \mathcal{L}(V, W) \to M_{m \times n}(\mathbb{F})$ be defined by $\phi(T) = [T]_{\mathcal{B}\to\mathcal{B}'}$. Then ϕ is a linear transformation.

Proof: Let $T, S \in \mathcal{L}(V, W)$. For every $\mathbf{b}_i \in \mathcal{B}$, $(cT + S)(\mathbf{b}_i) = cT(\mathbf{b}_i) + S(\mathbf{b}_i)$, therefore,

$$[(cT + S)(\mathbf{b}_i)]_{\mathcal{B}'} = [cT(\mathbf{b}_i) + S(\mathbf{b}_i)]_{\mathcal{B}'} = c[T(\mathbf{b}_i)]_{\mathcal{B}'} + [S(\mathbf{b}_i)]_{\mathcal{B}'}$$

Thus,

$$[cT + S]_{\mathcal{B}\to\mathcal{B}'} = c[T]_{\mathcal{B}\to\mathcal{B}'} + [S]_{\mathcal{B}\to\mathcal{B}'},$$

hence ϕ is a linear transformation.

Definition 2.4.1

A vector space A over a field $\mathbb{F}$ is said to be an associative algebra over $\mathbb{F}$ if in addition to the vector space operations, there is a function $\phi : A \times A \to A$ named multiplication such that $\phi(a, b) = ab$ satisfies:

1. $(ab)c = a(bc)$ for all $a, b, c \in A$, (multiplication associative);
2. $(a + b)c = ac + bc$ and $a(b + c) = ab + ac$ for all $a, b, c \in A$, (right and left distribution);
3. $(ka)\mathrm{b} = a(kb) = k(ab)$ for all a, $\mathrm{b} \in A$ and $k \in F$.

If there exists an element $1 \in A$ such that $1a = a1 = a$ for all $a \in A$, then A is an **algebra with multiplicative identity.**

Since the set of all linear transformation from V into V is a vector space over the field $\mathbb{F}$, we have that $\mathcal{L}(V, V)$ is an algebra with identity over $\mathbb{F}$. We note that the multiplication of two linear transformation is in fact the composition of these transformations. The multiplicative identity is the identity transformation.

Definition 2.4.2

Let A and B be algebra over $\mathbb{F}$. An algebra homomorphism from A to B is a linear transformation $\phi : A \to B$ such that $\phi(ab) = \phi(a)\phi(b)$ for all $a, b \in A$. An **algebra isomorphism** from A to B is a homomorphism which is bijective. Moreover, if $\phi : A \to B$ is an isomorphism, then A and B are isomorphic algebras.

Definition 2.4.3

The function T from V into W is called invertible if there exists a function U from W into V such that UT is identity function on V and TU is identity function on W. If T is invertible, the function U is unique and is denoted by T^{-1}. T is invertible if and only if

1. T is 1–1, that is $T(\alpha) = T(\beta)$ implies $\alpha = \beta$, and
2. T is onto, that is, the range of T is all of W.

By the Definition of invertible linear transformation, we have that

Theorem 2.4.2

Let V and W be vector spaces over the field $\mathbb{F}$ and let T be a linear transformation from V into W. If T is invertible, then the inverse function T^{-1} is a linear transformation from W onto V.

Moreover, let V and W be finite-dimensional vector spaces over the field $\mathbb{F}$ such that dim V = dim W. If T is a linear transformation from V into W, then the following are equivalent:

1. T is invertible;
2. T is non-singular;
3. T is onto, that is, the range of T is W;
4. If $\{\alpha_1, \ldots, \alpha_n\}$ is a basis for V, then $\{T(\alpha_1), \ldots, T(\alpha_n)\}$ is a basis for W;
5. For any basis $\{\beta_1, \ldots, \beta_n\}$ for W, there exists some basis $\{\alpha_1, \ldots, \alpha_n\}$ for V such that $T(\alpha_i) = \beta_i$ for $i = 1, \ldots, n$.

Definition 2.4.4

If V and W are vector spaces over the field $\mathbb{F}$, any invertible linear transformation $T : V \to W$ is called an **isomorphism of** V onto W. If there exists an isomorphism of V onto W, we say that V is **isomorphic** to W.

Definition 2.4.5

A **group** consists of the following:

1. A set G;
2. A rule or operation which associates with each pair of elements $x, y \in G$ an element $xy \in G$ in such a way that
   a. $x(yz) = (xy)z$, for all $x, y, z \in G$ (associativity);
   b. there is an element $e \in G$ such that $ex = xe = x$ for every $x \in G$;
   c. to each element $x \in G$ there corresponds an element $x^{-1} \in G$ such that $xx^{-1} = x^{-1}x = e$.

A group is called **commutative** if it satisfies the condition $xy = yx$ for each $x, y \in G$.

We see that the collection of invertible linear transformations in $\mathcal{L}(V, V)$, denoted as $GL(V)$ forms a group, called general linear group on V. Moreover, since any element of $\mathcal{L}(V, V)$, is isomorphic to a square matrix, the invertible linear transformation corresponds an invertible square matrix. The general linear group is a non-commutative group.

2.5 Applications of Linear Transformation

All linear transformations are formed by combining simple geometric processes such as rotation, stretching, shrinking, shearing, and projection. As a consequence linear transformations are important in computer graphics. For instance, to represent a 3-dimensional object on a 2-dimensional computer screen, or to look at the object from various angles, projections and rotations come into play. In addition, motion effects can be achieved by simply shrinking an object to make it appear to move away from the viewer. Linear transformations and matrices are also an important tool if we want to control the motion of a robot.

This geometric point of view is obviously useful when we want to model the motion or changes in shape of an object moving in the plane or in 3-space. However, this can be extended to higher dimensions as well. The idea that any matrix can be thought of as the product of simpler matrices that correspond to higher-dimensional versions of rotation, reflection, projection, shearing, dilation, and contraction is of enormous importance to both pure and applied mathematicians.

2.5.1 Affine Transformations

In mathematics, affine geometry is Euclidean geometry without the metric notions of distance and angle. Playfair's axiom (i.e., given a line L and a point P not on L, there is exactly one line parallel to L that passes through P) is fundamental in affine geometry. Affine geometry is developed on the basis of linear algebra. An affine space is a set of points equipped with a set of transformations called affine transformations, functions between affine spaces that preserve points, straight lines, and parallelism of lines.

Definition 2.5.1

Let X and Y be two affine spaces, then every affine transformation $T: X \to Y$ is of the form $T(\mathbf{x}) = M\mathbf{x} + \mathbf{b}$, where M is a linear transformation on X and $\mathbf{b}$ is a vector in Y.

An affine transformation does not necessarily preserve angles between lines or distances between points, though it does preserve ratios of distances between points lying on a straight line. Examples of affine transformations include translation, scaling, similarity transformation, reflection, rotation, shear mapping, and compositions of them in any combination and sequence. Unlike a linear transformation, an affine transformation need not preserve the zero point in a linear space. Thus, every linear transformation is affine, but not every affine transformation is linear.

In affine 3-space, a rigid motion, a transformation consisting of rotations and translations, is a transformation that when acting on any point p, generates a transformed point $T(p) = Rp + \mathbf{t}$, where R is a 3×3 orthogonal matrix (i.e., the columns of the matrix are pairwise orthogonal, and each column is of unit length) representing the rotation with $\det(R) = 1$, and $\mathbf{t} \in \mathbb{R}^3$ a 3-dimensional translation vector.

For any rotation matrix R acting on $\mathbb{R}^n$, $R^T = R^{-1}$, the rotation is an orthogonal matrix, and thus $\det R = \pm 1$. Since the inverse of a rotation matrix is its transpose, also a rotation matrix. The product of two rotation matrices is a rotation matrix. Multiplication of $n \times n$ rotation matrices is not commutative for $n > 2$. Moreover, any identity matrix is a rotation matrix, and that matrix multiplication is associative. Thus, the rotation matrices for $n > 2$ form a special non-commutative orthogonal group, and denoted by $SO(n)$, the group of $n \times n$ rotation matrices. Multiplication of rotation matrices corresponds to composition of rotations, applied in left-to-right order of their corresponding matrices.

Convert all points in 3-space to homogeneous coordinates:

$$\begin{bmatrix} x \\ y \\ z \end{bmatrix} \rightarrow \begin{bmatrix} x \\ y \\ z \\ 1 \end{bmatrix}$$

The following matrices constitute the basic affine transformations in 3-space, expressed in homogeneous form:

1. Translate: $\begin{bmatrix} 1 & 0 & 0 & \Delta x \\ 0 & 1 & 0 & \Delta y \\ 0 & 0 & 1 & \Delta z \\ 0 & 0 & 0 & 1 \end{bmatrix}$

2. Scale: $\begin{bmatrix} s_x & 0 & 0 & 0 \\ 0 & s_y & 0 & 0 \\ 0 & 0 & s_z & 0 \\ 0 & 0 & 0 & 1 \end{bmatrix}$

3. Shear: $\begin{bmatrix} 1 & h_{xy} & h_{xz} & 0 \\ h_{yx} & 1 & h_{yz} & 0 \\ h_{zx} & h_{zy} & 1 & 0 \\ 0 & 0 & 0 & 1 \end{bmatrix}$

4. Three basic rotations in 3-space:

Rotation about the x-axis: $\begin{bmatrix} 1 & 0 & 0 & 0 \\ 0 & \cos\theta_x & -\sin\theta_x & 0 \\ 0 & \sin\theta_x & \cos\theta_x & 0 \\ 0 & 0 & 0 & 1 \end{bmatrix}$

Rotation about the y-axis: $\begin{bmatrix} \cos\theta_y & 0 & \sin\theta_y & 0 \\ 0 & 1 & 0 & 0 \\ -\sin\theta_y & 0 & \cos\theta_y & 0 \\ 0 & 0 & 0 & 1 \end{bmatrix}$

Rotation about the z-axis: $\begin{bmatrix} \cos\theta_z & -\sin\theta_z & 0 & 0 \\ \sin\theta_z & \cos\theta_z & 0 & 0 \\ 0 & 0 & 1 & 0 \\ 0 & 0 & 0 & 1 \end{bmatrix}$

The rotations determine an amount of rotation about each of the individual axes of the coordinate system. The angles θ_x, θ_y and θ_z of rotation about the three axes are called the *Euler angles*. An off axis rotation can be achieved by combining combining Euler angle rotations via matrix multiplication. Since affine translformations are not commutative, the order of rotation affects the end result.

2.5.2 Projective Transformations

The idea of a projective space relates to perspective, that is, the way a camera projects a 3D scene to a 2D image. All points that lie on a projection line through the entrance pupil of the camera are projected onto a common image point. Hence, the vector space is $\mathbb{R}^3$ with the origin being the camera entrance pupil, and the projective space corresponds to the image points.

The real n-dimensional projective space or projective n-space, $\mathbb{P}^n_{\mathbb{R}}$, is the set of the lines in $\mathbb{R}^{n+1}$ passing through the origin, or

$$\mathbb{P}^n_{\mathbb{R}} := \{\mathbb{R}^{n+1} \neq \mathbf{0} / \sim\}$$

where $(x_0, \ldots, x_n) \sim (y_0, \ldots, y_n)$ *if* $(x_0, \ldots, x_n) = \lambda(y_0, \ldots, y_n)$ for some $\lambda \neq 0$.

The elements of the projective space are commonly called points. The projective coordinates of a point $x = [x_0 : \ldots : x_n]$ where $(x_0, \ldots, x_n)$ is any element of the corresponding equivalence class. The colons and the brackets emphasizing that the right-hand side is an equivalence class, which is defined up to the multiplication by a non-zero constant.

Geometric objects, such as points, lines, or planes, can be given a representation as elements in projective spaces based on homogeneous coordinates. Transformations within and between projective spaces are called *projectivities* and are the fundamental concern of projective geometry. Certain properties such as collinearity, concurrency, tangency, and incidence remain invariant under the action of a projectivity, and they are referred to as projective invariants.

A projective transformation from $\mathbb{P}^n_{\mathbb{R}}$ to $\mathbb{P}^n_{\mathbb{R}}$ is an invertible $(n + 1) \times (n + 1)$ matrix P. P acts on the projective n-space as the following:

$$Px = P\begin{bmatrix} x_0 \\ \vdots \\ x_n \end{bmatrix} = \begin{bmatrix} y_0 \\ \vdots \\ y_n \end{bmatrix} = y, \text{ where } P \text{ is an invertible } (n+1) \times (n+1) \text{ matrix.}$$

$\mathbf{x}$ and $\mathbf{y}$ are homogeneous coordinates for a point in a projective n-space.

Lemma Let P be a real projective transformation represented by a nonsingular $(n + 1) \times (n + 1)$ matrix $p = (P_{i,j})_{i,j=0}^{n}$. Then the following assertions are equivalent:

1. The restriction of P to $\mathbb{A}^n = \{[x_0 : \ldots : x_{n-1} : 1] \in \mathbb{P}^n\}$ is an affine transformation;
2. $P_{0,n} \ldots P_{(n-1),n} = 0$;
3. P fixes the hyper-plane $x_n = 0$ at infinity.

Proof: $1 \Leftrightarrow 2$: If

$$P\begin{bmatrix} x_0 \\ \vdots \\ x_{n-1} \\ 1 \end{bmatrix} = \begin{bmatrix} y_0 \\ \vdots \\ y_{n-1} \\ y_n \end{bmatrix} \text{ then } y_n = \sum_{i=0}^{n-1} P_{n,i}x_i + P_{n,n}$$

If P induces an affine transformation, then we must have $y_n \neq 0$ for all $x_i \in \mathbb{R}$, and this implies $P_{0,n} = \cdots = P_{(n-1),n} = 0$. Since $\det P \neq 0$, we must have $P_{n,n} \neq 0$. Thus, by rescaling, $P_{n,n} = 1$.

Conversely, if $P_{0,n} = \cdots = P_{(n-1),n} = 0$ and $P_{n,n} = 1$, then

$$P\begin{bmatrix} x_0 \\ \vdots \\ x_{n-1} \\ 1 \end{bmatrix} = \begin{bmatrix} y_0 \\ \vdots \\ y_{n-1} \\ 1 \end{bmatrix}, \text{ and this is an affine transformation.}$$

$2 \Leftrightarrow 3$: If $P_{0,n} = \ldots = P_{(n-1),n} = 0$, then

$$P\begin{bmatrix} x_0 \\ \vdots \\ x_{n-1} \\ 0 \end{bmatrix} = \begin{bmatrix} y_0 \\ \vdots \\ y_{n-1} \\ 0 \end{bmatrix}, \text{ hence the hyper-plane } x_n = 0 \text{ is preserved.}$$

Conversely, if

$$P\begin{bmatrix} x_0 \\ \vdots \\ x_{n-1} \\ 0 \end{bmatrix} = \begin{bmatrix} y_0 \\ \vdots \\ y_{n-1} \\ 0 \end{bmatrix}, \forall y_0, \cdots, y_{n-1} \in \mathbb{R},$$

then $P_{0,n} = \ldots = P_{(n-1),n} = 0$.

Proposition Let $P_i = [x_i : y_i : Z_i]$, for $i = 1, 2, 3, 4$ be four points in the projective plane, no three of which are collinear. Then there is a unique projective transformation sending the standard frame, namely $[1 : 0 : 0]$, $[0 : 1 : 0]$, $[0 : 0 : 1]$ and $[1 : 1 : 1]$ to P_1, P_2, P_3 and P_4.

Proof: Since P_i for $i = 1, 2, 3$ are non-collinear, we must have that

$$\det\begin{bmatrix} x_1 & x_2 & x_3 \\ y_1 & y_2 & y_3 \\ z_1 & z_2 & z_3 \end{bmatrix} \neq 0$$

In projective space

$[1 : 0 : 0] = [a_1 : 0 : 0]$, $[0 : 1 : 0] = [0 : a_2 : 0]$, $[0 : 0 : 1] = [0 : 0 : a_3]$, $a_i \neq 0$. Hence, there exists a 3×3 matrix P such that

$$P\begin{bmatrix} a & 0 & 0 \\ 0 & b & 0 \\ 0 & 0 & c \end{bmatrix} = \begin{bmatrix} x_1 & x_2 & x_3 \\ y_1 & y_2 & y_3 \\ z_1 & z_2 & z_3 \end{bmatrix} \Rightarrow \begin{bmatrix} x_1 & x_2 & x_3 \\ y_1 & y_2 & y_3 \\ z_1 & z_2 & z_3 \end{bmatrix}\begin{bmatrix} \frac{1}{a_1} & 0 & 0 \\ 0 & \frac{1}{a_2} & 0 \\ 0 & 0 & \frac{1}{a_3} \end{bmatrix}$$

Hence P is an invertible matrix. Hence this is a projective transformation.

Now P_4 will be the image of $[1 : 1 : 1]$ if and only if $P_4 = P[1, 1, 1]^T$.

Hence, $P_4 = \left[\sum_{i=1}^{3} \frac{x_i}{a_i} : \sum_{i=1}^{3} \frac{y_i}{a_i} : \sum_{i=1}^{3} \frac{z_i}{a_i}\right]^T$. One can check P_4 is linearly independent from any two of P_1, P_2, P_3.

Corollary: Let P_i and Q_i for $i = 1, 2, 3, 4$ be two sets of four points in the projective plane such that no three P_i and no three Q_i are collinear. Then there is a projective transformation sending P_i to Q_i for $i = 1, 2, 3, 4$.

Proof. Let P denote the projective transformation that sends the standard frame to the P_i, and let Q denote the transformation that does the same with the Q_i. Then QP^{-1} is the projective transformation sending P_i *to* Q_i.

Proposition An invertible projective transformation preserves the degree of curves in projective space.

Proof: An invertible projective transformation maps a monomial $x^i y^j z^k$ of degree $m = i + j + k$ either to 0 or to another homogeneous polynomial of degree m. We claim the case that the image being zero is not possible. Let P be a projective transformation, and $f(x, y, z)$ is a curve of degree m. If $T(f(x, y, z)) = 0$, then $f(x, y, z) = T^{-1}(0) = 0$ since T is invertible. It is not possible. Hence, the projective transformations preserve the degree of curves.

2.6 Exercises

1. Let V be a vector space and $f: V \to R$ be a linear map. *If* $z \in V$ is not in the null space of f, show that every $x \in V$ can be decomposed uniquely as $x = v + cz$, where v is in the null space of f and c is a scalar.

2. Let $V = \mathbb{R}^2$ and $W = \mathbb{R}^3$. Define $L: V \to W$ by $L(x, y) = (x - y, x, y)$.Let $F = \{(1, 1), (-1, 1)\}$, $G = \{(1, 0, 1), (0, 1, 1), (1, 1, 0)\}$. (a) Find the matrix representation of L using the standard bases in both V and W. (b) Find the matrix representation of L using the standard basis in V and the basis G in W. (c) Find the matrix representation of L using the basis F in $\mathbb{R}^2$ and the standard basis in $\mathbb{R}^3$. (d) Find the matrix representation of L using the bases F and G.

   Remark Here we include a solution provided by Andrew Crutcher, a student of this class. We can utilize the following code to find the matrix representation of an arbitrary T from B_1 to B_2 or $[T]_{B_1 \mapsto B_2}$. With this code we can take care of parts a, b, c, and d easily.

```
[style=pyStyle]
def matrix_rep(T,B1,B2):
  A = matrix.zero(QQ,T.nrows(),T.ncols())
  areBasis = B1.is_invertible() and B2.is_invertible()
  if areBasis and T.ncols() == B1.ncols() and
     T.nrows() == B2.nrows():
     B2inv = B2.inverse()
     for colNum in xrange(0,T.ncols()):
      A.set_column(colNum,B2inv*T*B1.column(colNum))
  return A
L=matrix(QQ,3,2, [1,-1,1,0,0,1])
F=matrix(QQ,2,2, [1,1,-1,1])
G=matrix(QQ,3,3, [1,0,1,0,1,1,1,1,0])
E22=matrix.identity(2)
E33=matrix.identity(3)
print("a) L from standard to standard")
matrix_rep(L,E22,E33)
print("b) L from standard to G")
matrix_rep(L,E22,G)
print("c) L from F to standard")
matrix_rep(L,F,E33)
print("d) L from F to G")
matrix_rep(L,F,G)
```

```
This results in the following output.
[style=pyStyle]
(a) L from standard to standard
[ 1 -1]
[ 1  0]
[ 0 1]
(b) L from standard to G
[ 0  0]
[ 0  1]
[ 1 -1]
(c) L from F to standard
[ 2 0]
[ 1 1]
[-1 1]
(d) L from F to G
[ 0 0]
[-1 1]
[ 2 0]
```

3. Let $B = \{\mathbf{b}_1, \ldots, \mathbf{b}_n\}$ be a basis for $\mathbb{R}^n$ and let $T : \mathbb{R}^n \to \mathbb{R}^n$ be defined as follows.

$$T\left(\sum_{k=1}^{n} a_k \mathbf{b}_k\right) = \sum_{k=1}^{n} a_k b_k \mathbf{b}_k$$

First show that T is a linear transformation. Next show that the matrix of T with respect to this basis, $[T]_B$ is

$$\begin{bmatrix} b_1 & & \\ & \ddots & \\ & & b_n \end{bmatrix}$$

If $E = \{\mathbf{e}_1, \ldots, \mathbf{e}_n\}$ be the standard basis for $\mathbb{R}^n$. Show that

$$[T]_E = (\mathbf{b}_1 \ldots, \mathbf{b}_n)[T]_B(\mathbf{b}_1 \ldots, \mathbf{b}_n)^{-1}$$

4. If V be the vector space of 2×2 matrices and $KM = \begin{bmatrix} 1 & 2 \\ 0 & 3 \end{bmatrix}$. If $T: V \to V$ be a linear transformation defined by $T(A) = AM - MA$ for every $A \in V$. Then what is the dimension of kernel of T?

5. Let P be the vector space of all polynomial functions on $\mathbb{R}$ with real coefficients. Define linear transformations $T, D : P \to P$ by $(Dp)(x) = p'(x)$ and $(Tp)(x) = x^2 p(x)$ for all $x \in \mathbb{R}$. Find matrix representations for the linear transformations $D + T$, DT, and TD (with respect to the usual basis $\{1; x; x^2\}$ for the space of polynomials of degree two or less).

6. Let A is an $n \times n$ matrix over $\mathbb{R}$ of rank k. Define

$$L = \{B\ n \times n \text{ matrix over } \mathbb{R} \mid BA = 0\}$$

and

$$R = \{Cn \times n \text{ matrix over } \mathbb{R} \mid AC = 0\}.$$

Show that L and R are linear spaces and compute their dimensions.

7. Find the null space of the matrix $\begin{bmatrix} 1 & 2 & 3 & 4 & 5 \\ 2 & 3 & 4 & 5 & 6 \\ 1 & 0 & 2 & 0 & 6 \end{bmatrix}$

8. Let $p_1 = x^3 + x + 1$, $p_2 = 2x^3 - x^2$, $p_3 = x^2 + x + 1$, $q_1 = x^3 - 1$, $q_2 = x^3 - x^2 + 1$, $q_3 = x^3 + x$ be polynomials. Is it true that Span $\{p_1, p_2, p_3\}$ = Span$\{q_1, q_2, q_3\}$?

9. Find the kernel of the linear transformation from $\mathbb{R}^4 \to \mathbb{R}^2$ with the following matrix: $\begin{bmatrix} 1 & 2 & 3 & 4 \\ 2 & 3 & 4 & 5 \end{bmatrix}$. Is the range of the transformation the same as $\mathbb{R}^2$?

10. What is the dimension of the subspace of $\mathbb{R}^4$ spanned by the column vectors

$$(1, 2, 3, 4), (1, 1, 1, 1), (3, 4, 5, 6), (5, 7, 9, 11).$$

Also find the kernel of the linear transformation from $\mathbb{R}^4 \to \mathbb{R}^4$ with the matrix formed by the above column vectors.

11. Is the function $T : M_{2\times 2} \to P_2$ defined by

$$T\begin{bmatrix} a & b \\ c & d \end{bmatrix} = ax^2 + bx + c(c+d)^2$$

a linear transformation?

12. Compute the dimension of the range of T where $T : \mathbb{R}^3 \to P_2$ is the linear transformation

$$T\begin{bmatrix} a \\ b \\ c \end{bmatrix} = (a+b)x^2 + (a+c).$$

13. Let $S = \{\mathbf{e}_1, \mathbf{e}_2\}$ denote the standard ordered basis for $\mathbb{R}^2$ and let $B = \{x - 1; x + 2\}$ be a basis for P_1. Let $T : \mathbb{R}^2 \to P_1$ be the linear a transformation $T\begin{bmatrix} a \\ b \end{bmatrix} = ax + 2(a+b)$. Find $[T]_{S \to B}$.

14. Is the linear transformation $T : P_2 \to \mathbb{R}^3$ given by $T(ax^2 + bx + c) = \begin{bmatrix} a+b \\ ac \\ b+c \end{bmatrix}$ an isomorphism of P_2 onto $\mathbb{R}^3$?

15. Let A be a linear transformation from $\mathbb{R}^6 \to \mathbb{R}^4$ find the dimensions of the kernel and range of A, where

$$A\begin{bmatrix} 1 & 0 & -3 & 0 & 2 & -8 \\ 0 & 1 & 5 & 0 & -1 & 4 \\ 0 & 0 & 0 & 1 & 7 & -9 \\ 0 & 0 & 0 & 0 & 0 & 0 \end{bmatrix}$$

Also identify a basis for the column space and a basis for the null space.

16. Let $f : \mathbb{R} \to \mathbb{R}$ be defined by $f(x) = 2x - 3$. Show f is one-to-one and onto; find an inverse map f^{-1}.

17. Suppose a linear mapping $F : V \to U$ is one-to-one and onto. Show that the inverse mapping $F^1 : U \to V$ is also linear.

18. Consider the linear operator T on $\mathbb{R}^3$ defined by

$$T(x, y, z) = (2x;\ 4x - y;\ 2x + 3y - z).$$

Show T is invertible, find formulas for T^{-1}, T^2, and T^{-2}.

Remark Below, we include a solution provided by Andrew Crutcher via computer software. Note to show T is invertible all we have to do is show $\ker(T) = \{\mathbf{0}\}$.

$$\text{Let } T(x, y, z) = 0 \Rightarrow \begin{matrix} 2x = 0 \\ 4x - y = 0 \\ 2x + 3y - z = 0 \end{matrix} \Rightarrow \begin{matrix} x = 0 \\ y = 0 \\ 3y - z = 0 \end{matrix} \Rightarrow \begin{matrix} x = 0 \\ y = 0 \\ z = 0 \end{matrix}$$

$\Rightarrow \ker(T) = \{\mathbf{0}\}$

$\therefore$ T is invertible.

Knowing the matrix for T is $\begin{bmatrix} 2 & 0 & 0 \\ 4 & -1 & - \\ 2 & 3 & -1 \end{bmatrix}$ allows us to utilize SAGE *to* calculate T^{-1}, T^2, T^{-2}

```
[style=pyStyle]
T = matrix(QQ,3,3,[2,0,0,4,-1,0,2,3,-1])
print("T"); T
```

```
print("T^-1"); T.inverse()
print("T^2"); (T)^2
print("T^-2"); (T.inverse())^2

This results in the following output.
[style=pyStyle]
T
[ 2 0  0]
[ 4 -1 0]
[ 2 3 -1]
T^-1
[1/2 0  0]
[ 2 -1  0]
[ 7 -3 -1]
T^2
[ 4  0 0]
[ 4  1 0]
[14 -6 1]
T^-2
[ 1/4  0 0]
[ -1   1 0]
[-19/2 6 1]
```

19. The set $S = \{e^{3t}, te^{3t}, t^2e^{3t}\}$ is a basis of a vector space V of functions $f: \mathbb{R} \to \mathbb{R}$. Let D be the differential operator on V; that is, $Df = df/dt$. Find the matrix representation of D relative to the basis S.

20. Suppose that the x-axis and y-axis in the plane $\mathbb{R}^2$ are rotated counterclockwise 30° to yield new X-axis and Y-axis for the plane. Find (a) The unit vectors in the direction of the new X-axis and Y-axis. (b) The change-of-basis matrix P for the new coordinate system. (c) The new coordinates of the points $A = (1, 3)$, $B = (2, 5)$, and $C = (a, b)$.

21. Let $T: \mathbb{R}^3 \to \mathbb{R}^2$ be defined by $T(x, y, z) = (2x + y - z, 3x - 2y + 4z)$. (a) Find the matrix A representing T relative to the bases $S = \{(1, 1, 1), (1, 1, 0), (1, 0, 0)\}$, and $S' = \{(1, 3), (1, 4)\}$. (b) Verify that, for any $\mathbf{v} \in \mathbb{R}^3$, $A[\mathbf{v}]_S = [T\mathbf{v}]_{S'}$.

22. Determine whether or not each of the following linear maps is nonsingular. If not, find a non-zero vector $\mathbf{v}$ whose image is 0. (a) $F: \mathbb{R}^2 \to \mathbb{R}^2$ defined by $F(x, y) = (x - y, x + 2y)$. (b) $G: \mathbb{R}^2 \to \mathbb{R}^2$ defined by $G(x, y) = (2x + 4y, 3x + 6y)$.

CHAPTER 3

Linear Operators

3.1 Characteristic and Minimal Polynomials

Definition 3.1.1

Let T be a linear transformation represented by a matrix A. If there is a vector $\mathbf{x}$ in $V \neq \mathbf{0}$ such that

$$A\mathbf{x} = \lambda\mathbf{x}, \text{ for some } \lambda \in \mathbb{F},$$

then λ is called the eigenvalue of A with corresponding eigenvector $\mathbf{x}$.

If A is a $k \times k$ square matrix $A = (a_{ij})$ with an eigenvalue λ, then the corresponding eigenvectors satisfy

$$A\mathbf{x} - \lambda\mathbf{x} = (A - \lambda\mathbb{I})\mathbf{x} = \mathbf{0}$$

As shown in Cramer's rule, a linear system of homogenous equations has non-trivial solutions if and only if the determinant vanishes, so the condition for existence of the solutions is given by

$$\det(A - \lambda\mathbb{I}) = 0.$$

This equation is known as the **characteristic equation of** A, and the left-hand side is known as the **characteristic polynomial.**

Therefore, we have the following theorem:

Theorem 3.1.1

Let T be a linear operator on a finite-dimensional space V and let c be a scalar. The following are equivalent:

(a) c is an eigenvalue of T;

(b) The operator $(T - c\mathbb{I})$ is singular;

(c) $\det(T - c\mathbb{I}) = 0$.

Proof: The proof directly follows the definition. The scalar c is an eigenvalue of T if and only if $T\mathbf{v} = c\mathbf{v}$ for some non-zero vector $\mathbf{v}$ if and only if $(T - c\mathbb{I})\mathbf{v} = \mathbf{0}$ if and only if $\det(T - c\mathbb{I}) = 0$, i.e., $(T - c\mathbb{I})$ is singular.

***Lemma*:** Suppose $T(\mathbf{v}) = c\mathbf{v}$. If f is any polynomial, then $(f(T))(\mathbf{v}) = (f(c))(\mathbf{v})$.

Proof: Let $f(x) = a_n x^n + a_{n-1}x^{n-1} + \cdots + a_1 x + a_0$. Then

$$\begin{aligned}(f(T))(\mathbf{v}) &= (a_n T^n + a_{n-1}T^{n-1} + \cdots + a_1 T + a_0\mathbb{I})(\mathbf{v})\\ &= a_n(c^n\mathbf{v}) + a_{n-1}(c^{n-1}\mathbf{v}) + \cdots + a_1(c\mathbf{v}) + a_0\mathbf{v}\\ &= (a_n c^n + a_{n-1}c^{n-1} + \cdots + a_1 c + a_0)\mathbf{v}\\ &= (f(c))(\mathbf{v})\end{aligned}$$

Theorem 3.1.2

Every operator on a finite-dimensional, non-zero, complex vector space has an eigenvalue.

Proof: Suppose V is a complex vector space with dimension $n > 0$ and T a linear operator. Let $\mathbf{0} \neq \mathbf{v} \in V$. Then

$$\mathbf{v}, T\mathbf{v}, T^2\mathbf{v}, \ldots, T^n\mathbf{v}$$

is not linearly independent, because $\dim V = n$ and we have $n + 1$ vectors. Thus there exist complex numbers $a_o, \ldots, a_n$, not all 0. Assume $a_n \neq 0$, such that

$$a_0\mathbf{v} + a1T\mathbf{v} + \ldots + a_n T^n\mathbf{v} = \mathbf{0}.$$

In case $a_n = 0$, then go to the next highest non-zero a_k, note that $k \geq 1$. Make the a_i's the coefficients of a polynomial, which by the fundamental theorem of algebra has a factorization

$$a_0 + a_1 x + \cdots + a_n x^n = c(x - \lambda_1) \ldots (x - \lambda_n)$$

where c is a non-zero complex number, each $\lambda_i \in \mathbb{C}$, and the equation holds for all $x \in \mathbb{C}$. We then have

$$a_0\mathbf{v} + a_1 T\mathbf{v} + \cdots + a_n T^n\mathbf{v} = c(T - \lambda_1\mathbb{I}) \ldots (T - \lambda_n\mathbb{I})(\mathbf{v})$$

Thus $\lambda_1, \ldots, \lambda_n$ are eigenvalues of T.

Moreover, we see that two matrices A, B are similar, then $A = P^{-1}BP$ for some invertible matrix P, then

$$\det(A - \lambda\mathbb{I}) = \det(P^{-1}BP - \lambda\mathbb{I}) = \det(P^{-1}(B - \lambda\mathbb{I})P) = \det(B - \lambda\mathbb{I}).$$

Hence we conclude that similar matrices have the same characteristic polynomial.

The **algebraic multiplicity** $\mu_A(\lambda_i)$ of the eigenvalue is its multiplicity as a root of the characteristic polynomial, that is, the largest integer k_i such that $(\lambda - \lambda_i)^{k_i}$ divides evenly that polynomial.

Suppose a $k \times k$ matrix A has d, $d \leq k$, distinct eigenvalues. Hence the characteristic polynomial of A factors into the product of k linear terms over $\mathbb{C}$ with some terms potentially repeating, the characteristic polynomial can instead be written as the product d terms each corresponding to a distinct eigenvalue and raised to the power of the algebraic multiplicity,

$$\det(A - \lambda \mathbb{I}) = (\lambda - \lambda_1)^{k_1} \cdots (\lambda - \lambda_2)^{k_2} (\lambda - \lambda_d)^{k_d}, \sum_{i=1}^{d} k_i = k$$

The subspace generated by the eigenvectors corresponding to an eigenvalue λ is called an eigen-subspace associated with λ. The dimension of an eigen-subspace associated with an eigenvalue λ is called the **geometric multiplicity of the eigenvalue** λ, i.e. dim ker$(\lambda - \lambda\mathbb{I})$. It is a proven result that the geometric multiplicity is at most equal to the algebraic multiplicity of a particular eigenvalue.

Definition 3.1.2

Let T be a linear operator on the finite-dimensional space V. We say that T *is* **diagonalizable** if there is a basis for V each vector of which is a eigenvector of T.

The following corollary is an easy observation from the above definition.

***Corollary*:** Let $\mathcal{B}$ be a basis for a finite-dimensional vector space V such that $\mathcal{B}$ diagonalize T. Then $[T]_{\mathcal{B}}$ has all eigenvalues on the main diagonal. Moreover, the number of repetition of an eigenvalue is the algebraic multiplicity of that eigenvalue.

Definition 3.1.3

The companion matrix of the monic polynomial

$$p(t) = c_0 + c_1 t + \ldots + c_{n-1}t^{n-1} + t^n$$

is the square matrix defined as

$$C(p) = \begin{bmatrix} 0 & 0 & & 0 & \\ 1 & 0 & & 0 & \\ 0 & 1 & & 0 & \\ \vdots & \vdots & \ddots & \vdots & \vdots \\ 0 & 0 & & 1 & \end{bmatrix}, \quad c \in$$

With this convention, with standard basis $\mathcal{B} = \{\mathbf{e}_1, \ldots, \mathbf{e}_n\}$, one has

$$(C^T)(\mathbf{e}_i) = (C^T)^i(\mathbf{e}_1) = \mathbf{e}_{i+1}, \forall\, i < n,$$

and $\mathbf{e}_1$ generates the vector space V as a $\mathbb{F}[C^T]$-module over a field $\mathbb{F}$, and we call $\mathbf{e}_i$ are C^T cycles basis vectors.

A **Vandermonde matrix** is a $m \times n$ matrix

$$\mathcal{V} = \begin{bmatrix} 1 & \alpha_1 & \alpha_1^2 & \cdots & \alpha_1^{n-1} \\ 1 & \alpha_2 & \alpha_2^2 & \cdots & \alpha_2^{n-1} \\ 1 & \alpha_3 & \alpha_3^2 & \cdots & \alpha_3^{n-1} \\ \vdots & \vdots & \vdots & \ddots & \vdots \\ 1 & \alpha_m & \alpha_m^2 & \cdots & \alpha_m^{n-1} \end{bmatrix},$$

or

$$\mathcal{V}_{i,j} = \alpha_i^{j-1}, \forall i, j$$

The determinant of a square Vandermonde matrix can be expressed as

$$\det(\mathcal{V}) = \prod_{i \le i < j \le n} (\alpha_j - \alpha_i)$$

This is called the **Vandermonde determinant** or **Vandermonde polynomial**. If all the numbers ai are distinct, then it is non-zero.

As it turns out, the roots of the characteristic polynomial of a companion matrix C, i.e., $\det(C - x\mathbb{I}) = 0$ are precisely the roots of the polynomial $p(x)$. The eigenvector corresponding to the eigenvalue λ_j is Vandermonde vector $\mathbf{v} = [1, \lambda_j, \ldots, \lambda_j^{n-1}]^T$. Then, if α_j is a root of $p(x)$, and $\mathbf{v}$ is a Vandermonde vector such that $\mathbf{v} = [1, \alpha_j, \ldots, \alpha_j^{n-1}]^T$, then

$$C\mathbf{v} = \begin{bmatrix} 0 & 0 & \cdots & 0 & -c_0 \\ 1 & 0 & \cdots & 0 & -c_1 \\ 0 & 1 & \cdots & 0 & -c_2 \\ \vdots & \vdots & \ddots & \vdots & \vdots \\ 0 & 0 & \cdots & 1 & -c_{n-1} \end{bmatrix}^T \begin{bmatrix} 1 \\ \alpha_j \\ \vdots \\ \alpha_j^{n-1} \end{bmatrix} = \begin{bmatrix} \alpha_j \\ \vdots \\ \alpha_j^{n-1} \\ -\sum_{i=1}^{n-1} c_i \alpha_j^i \end{bmatrix} = \begin{bmatrix} \alpha_j \\ \vdots \\ \alpha_j^{n-1} \\ \alpha_j^n \end{bmatrix} = \alpha_j \mathbf{v}$$

Definition 3.1.4

Let $T : V \to V$ be a linear operator on a finite-dimensional vector space over the field $\mathbb{F}$. We shall denote by T^k the composition of T with itself k times, and for any polynomial

$$p(t) = akt^k + \cdots + a_0$$

we set

$$p(T) = ak\ T^k + \ldots + a_1T + a_0\mathbb{I};$$

and say *that* p *is* **monic** *if* $a_k = 1$. *A* **minimal polynomial** μ_T of the linear operator T is a monic polynomial of minimal degree such that $\mu_T(T) = 0$.

Let us recall that a polynomial ring $\mathbb{F}[x]$ is the subspace $\mathbb{F}^\infty$ spanned by the vectors $1, x, x^2, \ldots$. An element of $\mathbb{F}[x]$ is called a polynomial over $\mathbb{F}$. Let $\mathbb{F}$ be a field. An ideal in $\mathbb{F}[x]$ is a subspace $M \subseteq \mathbb{F}[x]$ such that $fg \in M$ whenever $f \in \mathbb{F}[x]$, and $g \in M$. The set $M = d\mathbb{F}[x]$, all multiples df of d by arbitrary $f \in \mathbb{F}[x]$ is an ideal. The ideal M is called the principal ideal generated by d. But if $M = \sum_{i=1}^{n} d_i\mathbb{F}[x]$, then M is generated by $d_1, \ldots, d_n$.

If T is a linear operator on a finite-dimensional vector space V over the field $\mathbb{F}$. Then the minimal polynomial for T is in fact the unique monic generator of the ideal of polynomials over $\mathbb{F}$ which annihilate T. Moreover, the characteristic and minimal polynomial for T have the same roots, except possibly for multiplicity. Hence we can summarize as below remark:

***Remark*:** Let $T : V \rightarrow V$ be a linear operator on a finite-dimensional vector space V over the field $\mathbb{F}$. Then:

(a) the minimal polynomial μ_T of T exists, has degree at most $n = \dim V$, and is unique;

(b) if $p \in \mathbb{F}[t]$ is such that $p(T) = 0$, then there is some $q \in \mathbb{F}[t]$ such that $p = q\mu_T$.

The minimal polynomial can be explicitly computed. Take $\mathbf{v} \in V$, and let d be the minimal non-negative integer such that the elements in $\{\mathbf{v}, T(\mathbf{v}), \ldots, T^d(\mathbf{v})\}$ are linearly dependent. Clearly $d \leq n$; $d = 0$ if and only if $\mathbf{v} = \mathbf{0}$, and $d = 1$ if and only if $\mathbf{v}$ is an eigenvector of T. Since elements in $\{\mathbf{v}, T(\mathbf{v}), \ldots, T^d(\mathbf{v})\}$ are linearly dependent, there exist $a_0, \ldots, a_{d-1} \in \mathbb{F}$ such that

$$T^d(\mathbf{v}) + a_{d-1}T^{d-1}(\mathbf{v}) + \cdots + a_1T(\mathbf{v}) + a_0\mathbf{v} = \mathbf{0}$$

Note that we can assume the coefficient of $T^d(\mathbf{v})$ is 1 because of the minimality of d, and then set

$$\mu\mathrm{T,v}(t) = t^d + a_{d-1}t^{d-1} + \cdots + a_1t + a_0.$$

By definition, $\mathbf{v} \in \ker \mu_T,\mathrm{v}(T)$; more precisely, $\mu_{T,\mathbf{v}}(T)$ is the monic polynomial $p(T)$ of least degree such that $\mathbf{v} \in \ker p(T)$. Now, if $p \in \mathbb{F}[t]$ is any common multiple of $\mu_{T,\mathbf{v}_1}$ **and** $\mu_{T,\mathbf{v}_2}$ for any two vectors $\mathbf{v}_1$ and $\mathbf{v}_2$, then both $\mathbf{v}_1$ and $\mathbf{v}_2$ belong to $\ker p(T)$. More generally, *if* $\mathcal{B} = \{\mathbf{v}_1, \ldots, \mathbf{v}_n\}$ is a basis of V, and p is any common multiple of $\mu_{T,\mathbf{v}_1}, \ldots, \mu_{T,\mathbf{v}_n}$, then $\mathcal{B} \subset \ker p(T)$, and $p(T) = 0$. We have the following result:

Theorem 3.1.3

Let $T : V \to V$ be a linear operator on a finite-dimensional vector space V over the field $\mathbb{F}$. Let $\mathcal{B} = \{\mathbf{v}_1, \dots, \mathbf{v}_n\}$ be a basis for V. Then μ_T is the least common multiple of $\mu_{T,\mathbf{v}_1}, \dots, \mu_{T,\mathbf{v}_n}$.

Proof: Let $p \in \mathbb{F}[t]$ be the least common multiple of $\mu_{T,\mathbf{v}_1}, \dots, \mu_{T,\mathbf{v}_n}$. Since $p(T) = 0$, and so $\mu_T \mid p$. Conversely, for $j = 1, \dots, n$, let

$$\mu_T = q_j \mu_{T,\mathbf{v}_j} + r_j, \text{ with } \deg r_j < \deg \mu_{t,\mathbf{v}_j}$$

Then

$$\mathbf{0} = \mu_t(T)\mathbf{v}_j = q_j(T)(\mathbf{v}_j)\mu_{T,\mathbf{v}_j}(T)(\mathbf{v}_j) + r_j(T)\mathbf{v}_j = r_j(T)\mathbf{v}_j;$$

and the minimality of the degree of $\mu_{t,\mathbf{v}_j}$ forces $r_j = 0$. Since every $\mu_{t,\mathbf{v}_j} \mid \mu_T$, their least common multiple $p \mid \mu_T$, and hence $p = \mu_T$.

In linear algebra, the Cayley-Hamilton theorem states that every square matrix over a commutative ring satisfies its own characteristic equation. When the ring is a field, the Cayley-Hamilton theorem is equivalent to the statement that the minimal polynomial of a square matrix divides its characteristic polynomial. Below we only state the theorem:

Theorem 3.1.4

(Cayley-Hamilton). Let T be a linear operator on a finite-dimensional vector space V. If f is the characteristic polynomial for T, then $f(T) = 0$; i.e., the minimal polynomial divides the characteristic polynomial. (Note, if the characteristic polynomial is $f = \prod_{i=1}^{k} (x - c_i)^{d_i}$, c_i distinct, $d_i \geq 1$, then the minimal polynomial $p = \prod_{i=1}^{k} (x - c_i)^{r_i}, 1 \leq r_i \leq d_i$).

3.2 Eigenspaces and Diagonalizability

For a given linear operator $T : V \to V$, a non-zero vector $\mathbf{v}$ and a constant scalar c are called an eigenvector and its eigenvalue, respectively, when $T\mathbf{v} = c\mathbf{v}$. For a given eigenvalue c, the set of all $\mathbf{v}$ such that $T\mathbf{v} = c\mathbf{v}$ is called the c-eigenspace. The set of all eigenvalues for an operator is called its **spectrum**. When the operator T is described by a matrix A, then we'll associate the eigenvectors, eigenvalues, eigenspaces, and spectrum to A as well.

Theorem 3.2.1

Each c-eigenspace is a subspace of V.

Proof: Suppose that **x** and **y** are c-eigenvectors and k is a scalar. Then

$$T(\mathbf{x}+k\mathbf{y}) = T(\mathbf{x}) + kT(\mathbf{y}) = c\mathbf{x} + kc\mathbf{y} = c(\mathbf{x}+ky),\ k \in \mathbb{F}.$$

Therefore $\mathbf{x} + k\mathbf{y}$ is also a c-eigenvector. Thus, the set of c-eigenvectors forms a subspace.

Definition 3.2.1

Let V be a vector space and T a linear operator on V. If W is a subspace of V, we say that W is invariant under T if for each vector $\mathbf{w} \in W$, the vector $T(\mathbf{w}) \in W$. That is $T(W) \subseteq W$.

Lemma: Let c_1 and c_2 be two distinct eigenvalues of a linear operator $T: V \to V$. Let W_i be the space of eigenvectors associated with the eigenvalue c_i. Then

(a) W_i is a subspace of V and it is invariant under T;

(b) $W_1 \cap W_2 = \{\mathbf{0}\}$.

Proof: First we note that $W_i = \ker(T - c_i\mathbb{I})$, hence it is a subspace. To prove this is invariant, let $\mathbf{v}, \mathbf{w} \in W_i$, then $T(a\mathbf{v} + b\mathbf{w}) = aT(\mathbf{v}) + bT(\mathbf{w}) = ac_1\mathbf{v} + bc_1\mathbf{w} = c_1(a\mathbf{v} + b\mathbf{w})$. Hence this is invariant under T.

To show the second claim, let $\mathbf{v} \in W_1 \cap W_2$, *then*

$T(\mathbf{v}) = c_1\mathbf{v}$, **and** $T(\mathbf{v}) = c_2\mathbf{v}, \Rightarrow c_1\mathbf{v} = c_2\mathbf{v} \Rightarrow (c_1 - c_2)\mathbf{v} = \mathbf{0}$, Þ $\mathbf{v} = \mathbf{0}$, since $c_1 = c_2$.

The above result shows that several eigenvectors correspond to a single eigenvalue, and an eigenvector cannot correspond to two (or more) distinct eigenvalues.

Hence, we have the following result:

Lemma 3.2.1: Let $T: V \to V$ be a linear operator on a finite-dimensional vector space V over the field $\mathbb{F}$. If T has distinct eigenvalues, then any eigenvectors corresponding to these eigenvalues respectively are linearly independent.

Proof: Assume the opposite, that is $\Sigma_{i=1}^{n}{}^{n} = a_i\mathbf{v}_i = \mathbf{0}$ for some $a_i \neq 0$, where $\mathbf{v}_i$ is the eigenvector corresponding to eigenvalue c_i. Without loss of generality, let us assume that $\mathbf{v}_1, \ldots, \mathbf{v}_{n-1}$ are linearly independent, and

$$\sum_{i=1}^{n} a_i\mathbf{v}_i = \mathbf{0} \quad \Rightarrow \quad a_n\mathbf{v}_n = -\sum_{i=1}^{n-1} a_i\mathbf{v}_i$$

Applying the linear operator T, we have

$$\mathbf{0} = T(\mathbf{0}) = T\left(\sum_{i=1}^{n} a_i \mathbf{v}_i\right) = \sum_{i=1}^{n} a_i c_i \mathbf{v}_i$$

and hence

$$\mathbf{0} = \sum_{i=1}^{n} a_i c_i \mathbf{v}_i = \sum_{i=1}^{n-1} a_i c_i \mathbf{v}_i + a_n c_n \mathbf{v}_n = \sum_{i=1}^{n-1} a_i c_i \mathbf{v}_i - \sum_{i=1}^{n-1} c_n a_i \mathbf{v}_i$$
$$= \sum_{i=1}^{n-1} a_i (c_i - c_n) \mathbf{v}_i$$

Since $\mathbf{v}_i$ are linearly independent, we must have $a_i(c_i - c_n) = 0$ for $i = 1, \ldots, n - 1$. Since not all a_i's are zero, $c_i - c_n$ for some i, a contradiction since eigenvalues are distinct.

***Corollary*:** Suppose V is finite-dimensional. Then each operator on V has at most dim V distinct eigenvalues.

Proof: Let T be a linear operator on V. Suppose $c_1, \ldots, c_n$ are distinct eigenvalues of T. Let $\mathbf{v}_1, \ldots, \mathbf{v}_n$ be corresponding eigenvectors. Then Lemma 3.2.1 implies that the list $\mathbf{v}_1, \ldots, \mathbf{v}_n$ is linearly independent. Thus $n \leq \dim V$ as desired.

Recall that the matrix of a linear map from one vector space to another vector space dependeds on a choice of a basis of each of the two vector spaces. Now that we are studying operators, which map a vector space to itself, the emphasis is on using only one basis. In this section, we consider the matrices of operators are square arrays, rather than the more general rectangular arrays that we considered earlier for linear maps.

A central goal of linear algebra is to show that given a linear operator T, there exists a basis of V with respect to which T has a reasonably simple matrix. To make this vague formulation a bit more precise, we might try to choose a basis of V such that the matrix representation of the linear operator has many zeros as possible.

Recall a matrix is called *upper triangular* if all the entries below the diagonal equal 0. Typically we represent an upper-triangular matrix in the form

$$\begin{bmatrix} \lambda_1 & & * \\ 0 & \ddots & \\ 0 & 0 & \lambda_n \end{bmatrix}$$

Upper triangular matrices can be considered reasonably simple for $n \gg 0$; almost half of its entries in an $n \times n$ upper triangular matrix are 0.

Theorem 3.2.2

Suppose T is a linear operator, and $\mathbf{v}_1, \ldots, \mathbf{v}_n$ is a basis for V. Then the following are equivalent:

1. The matrix of T with respect to $\mathbf{v}_1, \ldots, \mathbf{v}_n$ is upper triangular;
2. $T\mathbf{v}_j \in \text{Span}(\mathbf{v}_1, \ldots, \mathbf{v}_j)$ for each $j \in \{1, \ldots, n\}$;
3. $\text{Span}(\mathbf{v}_1, \ldots, \mathbf{v}_j)$ is invariant under T for each $j \in \{1, \ldots, n\}$.

Proof: The equivalence of the first two statements follows easily from the definitions.

Obviously (3) implies (2). Hence to complete the proof, we need only prove that (2) implies (3).

Thus suppose (2) holds. Fix $j \in \{1, \ldots, n\}$. From (2), we know that

$$T\mathbf{v}_1 \in \text{Span}(\mathbf{v}_1) \subset \text{Span}(\mathbf{v}_1, \ldots, \mathbf{v}_j)$$
$$T\mathbf{v}_2 \in \text{Span}(\mathbf{v}_1, \mathbf{v}_2) \subset \text{Span}(\mathbf{v}_1, \ldots, \mathbf{v}_j)$$
$$\vdots$$
$$T\mathbf{v}_j \in \text{Span}(\mathbf{v}1, \ldots, \mathbf{v}_j).$$

Thus if $\mathbf{v}$ is a linear combination of $\mathbf{v}_1, \ldots, \mathbf{v}_j$, then $T\mathbf{v} \in \text{Span}(\mathbf{v}_1, \ldots, \mathbf{v}_j)$. Hence $\text{Span}(\mathbf{v}_1, \ldots, \mathbf{v}_j)$ is invariant under T, completing the proof.

It is easy to see that the matrix of an operator in any basis $\mathcal{B}$ for V *is*

$$[T - \lambda\mathbb{I}]_{\mathcal{B}} = [T]_{\mathcal{B}} - \lambda[\mathbb{I}]_{\mathcal{B}}$$

If we now look for the coordinates of an eigenvector corresponding to the eigenvalue λ in basis $\mathcal{B}$, then

$$[T - \lambda\mathbb{I}]_{\mathcal{B}}[\mathbf{x}]_{\mathcal{B}} = ([T]_{\mathcal{B}} - \lambda[\mathbb{I}]_{\mathcal{B}})[\mathbf{x}]_{\mathcal{B}} = \mathbf{0}$$

Theorem 3.2.3

Suppose V is a finite-dimensional complex vector space and T is a linear operator on V. Then T has an upper-triangular matrix with respect to some basis of V.

Proof: We will use induction on the dimension of V. Clearly the desired result holds if $\dim V = 1$.

Suppose now that dim $V > 1$ and the desired result holds for all complex vector spaces whose dimension is less than dim V. Let c be any eigenvalue of T. Let

$$W = \text{range}(T - c\mathbb{I}).$$

Because $(T - c\mathbb{I})$ is not surjective dim $W <$ dim V. Furthermore, W is invariant under T, since for any $\mathbf{v} \in W$,

$$T\mathbf{v} = (T - c\mathbb{I})\mathbf{v} + c\mathbf{v} \in W, \text{ since } (T - c\mathbb{I})\mathbf{v} \in W, c\mathbf{v} \in W.$$

Thus T is an operator on W. By our induction hypothesis, there is a basis $\mathbf{v}_1$, ..., $\mathbf{v}_m$ of W with respect to which T has an upper-triangular matrix. Thus for each j we have

$$T\mathbf{v}_j \in \text{Span}(\mathbf{v}_1, \ldots, \mathbf{v}_m).$$

Then, we can extend $\mathbf{v}_1, \ldots, \mathbf{v}_m$ *to a basis* $\mathbf{v}_1, \ldots, \mathbf{v}_n$ *of* V. For $k = m + 1$, we have

$$Tv_{m+1} = (T - c\mathbb{I})\mathbf{v}_{m+1} + c\mathbf{v}_{m+1}.$$

Since the basis for range of $T - c\mathbb{I}$ is $\{\mathbf{v}_1, \ldots, \mathbf{v}_m\}$, we must have that $T\mathbf{v}_{m+1} \in \text{Span}(\mathbf{v}_1, \ldots, \mathbf{v}_m, \mathbf{v}_{m+1})$. Repeating this process, the equation above shows that $T\mathbf{v}_k \in \text{Span}(\mathbf{v}_1, \ldots, \mathbf{v}_k)$ for all $k = m + 1, \ldots, n$. Hence T has an upper triangular matrix with respect to the basis $\mathbf{v}_1, \ldots, \mathbf{v}_n$ of V.

Theorem 3.2.4

Suppose T is a linear operator has an upper-triangular matrix with respect to some basis of V. Then the eigenvalues of T are precisely the entries on the diagonal of that upper-triangular matrix.

Proof: Suppose $\mathcal{B} = \{\mathbf{v}_1, \ldots, \mathbf{v}_n\}$ is a basis for V with respect to which T has an upper-triangular matrix

$$[T]_{\mathcal{B}} = \begin{bmatrix} \lambda_1 & & * \\ 0 & \ddots & \\ 0 & 0 & \lambda_n \end{bmatrix}$$

Then

$$[T - \lambda\mathbb{I}]_{\mathcal{B}} = \begin{bmatrix} \lambda_1 - \lambda & & * \\ 0 & \ddots & \\ 0 & 0 & \lambda_n - \lambda \end{bmatrix}$$

Hence $[T - \lambda\mathbb{I}]_{\mathcal{B}}$ is not invertible if and only if λ equals one of the numbers λ_i for $i = 1, \ldots, n$. Thus $[T - \lambda\mathbb{I}]_{\mathcal{B}}$ is not invertible if and only if the kernel of

$(T - \lambda\mathbb{I})$ is non-trivial. Therefore, there exists a non-zero vector v such that $(T - \lambda\mathbb{I})\mathbf{v} = \mathbf{0}$, that is, $T\mathbf{v} = \lambda\mathbf{v}$. Thus A is an eigenvalue of T if and only if it equals one of the numbers λ_i for $i = 1, \ldots, n$.

We will see the relationship between the eigenvalues and the invariant subspaces. First, we will introduce the concept of vector space decomposition.

Definition 3.2.2

Let $W_1, \ldots, W_k$ be subspaces of the vector space V. We say that $W_1, \ldots, W_k$ are independent if

$$\mathbf{w}_1 + \ldots + \mathbf{w}_k = \mathbf{0},\ \mathbf{w}_i \in W_i \Rightarrow \mathbf{w}_i = \mathbf{0},\ \forall i.$$

Remark: Note this says that the

1. $\bigcap_{i=1}^{k} W_i = \{\mathbf{0}\}$;
2. $W_j \cap (W1 + \ldots + W_r) = \{0\}$ for all $r < j$, and for all j;
3. Each $\mathbf{w} \in \sum_{i=1}^{k} W_i, \mathbf{w} = \mathbf{w}_1 + \ldots + \mathbf{w}_k$ with $\mathbf{w}_i \in W_i$;
4. The expression $\mathbf{w} = \mathbf{w}1 + \ldots + \mathbf{w}_k$ with $\mathbf{w}_i \in W_i$ is unique.

Suppose $W_1, \ldots, W_k$ are independent subspaces of V where $V = \sum_{i=1}^{k} W_i$, and $\mathcal{B}_i$ is an ordered basis for W_i for $1 \le i \le k$. Then $\mathcal{B} = (\mathcal{B}_1, \ldots, \mathcal{B}_k)$ is an ordered basis for V. Let T be a linear operator on V, and $T_i = T|W_i$. Then T_i is a linear operator on W_i. *If* $A = [T]_{\mathcal{B}}$ and $A_i = [T_i]_{\mathcal{B}i}$, then A has the block form

$$A = \begin{bmatrix} A_1 & 0 & \cdots & 0 \\ 0 & A_2 & \cdots & 0 \\ \vdots & \vdots & & \vdots \\ 0 & 0 & \cdots & A \end{bmatrix}$$

Thus, A is a direct sum of the matrices $A_1, \ldots, A_k$.

The following lemma is a direct consequence from the definition of independent vector subspaces.

Lemma 3.2.2: Let V be a finite-dimensional vector space. Let $W_1, \ldots, W_k$ be subspaces of V, and $W = W_1 + \ldots + W_k$. The following are equivalent:

1. W_i are independent;
2. For each j, $2 \le j \le k$, we have $W_j \cap (W1 + \ldots + W_{j-1}) = \{\mathbf{0}\}$.
3. If $\mathcal{B}_i$ is an ordered basis for W_i, $1 \le i \le$ k, then the sequence $\mathcal{B} = (\mathcal{B}_1, \ldots, \mathcal{B}_k)$ is an ordered basis for W.

Definition 3.2.3

If any of the conditions of the above lemma holds, then the sum $W = W_1 + \cdots + W_k$ is direct or W is the direct sum of $W_1, \ldots, W_k$, and we write

$$W = W1 \oplus \cdots \oplus W_k.$$

Lemma 3.2.3: Let T be a linear operator on a finite-dimensional space V. Let $c_1, \ldots, c_k$ be the distinct eigenvalues of T, and let W_j be the space of eigenvectors associated with the eigenvalue c_j. If $W = W_1 + \cdots + W_k$, then

$$\dim W = \dim W_1 + \ldots + \dim W_k.$$

In fact, if B_i is an ordered basis for W_i, then $\mathcal{B} = (\mathcal{B}_1, \ldots, \mathcal{B}_k)$ is an ordered basis for W.

Proof: Since $W_i \cap W_j = \{\mathbf{0}\}$ for $i \neq j$, $W_1, \ldots, W_k$ are independent subspaces. Hence by definition, W is the direct sum of $W_1, \ldots, W_k$. Therefore,

$$W = W_1 \oplus W_2 \oplus \cdots \oplus W_k, \Rightarrow \dim W = \dim W_1 + \cdots + \dim W_k.$$

We can summarize that:

Theorem 3.2.5

Let T be a linear operator on a finite-dimensional space V. Let $c_1, \ldots, c_k$ be distinct eigenvalues of T and let W_j be the null space of $(T - c_j\mathbb{I})$. The following are equivalent:

1. T is diagonalizable;
2. The characteristic polynomial for T is $f = (x - c_1)^{d_1} \ldots (x - c_k)^{d_k}$, and $\dim W_i = d_j$, for $i = 1, \ldots, k$.
3. $W_1 + \cdots + W_k = W_1 \oplus \ldots \oplus W_k = V$, and $\dim W = \dim W_1 + \cdots + \dim W_k$.

Proof: $(1 \Rightarrow 3)$ Assume that T is diagonalizable. Then we can find a basis $\mathcal{B}$ for V consisting of eigenvectors for T. Each of these vectors is associated with a particular eigenvalue, so write $c_1, \ldots, c_k$ for the distinct ones. We can then group together the elements of $\mathcal{B}$ associated with c_j, span them, and call the resulting subspace W_j. It follows then that

$$W_1 + \cdots + W_k = W_1 \oplus \cdots \oplus W_k = \text{Span}\mathcal{B} = V.$$

$(3 \Rightarrow 2)$ Now assume that 3 holds. Then build a basis $\mathcal{B}_i$ of size n_i of each W_i. Since the c_i's are eigenvalues, the $\mathcal{B}_i$'s consist of eigenvectors for eigenvalue c_i and $n_i > 1$ for all i. Now, set $\mathcal{B} = (\mathcal{B}_1, \ldots, \mathcal{B}_k)$ is a basis for V and $[T]_{\mathcal{B}}$ is a diagonal matrix with distinct entries $c_1, \ldots, c_k$, with each c_i repeated n_i times. By computing the characteristic polynomial of the matrix $[T]_{\mathcal{B}}$, we find that item 2 holds.

$(2 \Rightarrow 1)$ Now if 2 holds, then $d_1 + \cdots + d_k = \dim V$, because the characteristic polynomial of $[T]_{\mathcal{B}}$ equals dim V. Therefore,

$$\dim V = dim\ W_1 + \cdots + \dim W_k.$$

Because each c_i is a root of the characteristic polynomial, it is an eigenvalue and therefore has an eigenvector. This means each dim $W_i \geq 1$. Let $\mathcal{B}_i$ be a basis for W_i. As the c_i's are distinct, the W_i's are independent subspaces, and so

$$W_1 + \cdots + W_k = W_i \oplus \cdots \oplus W_k.$$

This means $\mathcal{B} = (\mathcal{B}_1, \ldots, \mathcal{B}_k)$ is a basis for $W_1 \oplus \cdots \oplus W_k$. Since it has size of dim V, it is in fact a basis for V. But each vector of $\mathcal{B}$ is an eigenvector for T so T is diagonalizable.

Now, let us recall the minimal polynomial $\mu_T(x)$ *of T, and factor it completely into irreducible factors, to get* $\mu_T(x) = \prod_{i=1}^{k} (p_i(x))^r$. We then consider the linear transformations $(p_i(T))^{r_i}$, and their null spaces V_i. Each V_i is T-invariant, since if $\mathbf{v} \in V_i$, then $T(\mathbf{v}) \in V_i$. To see this, we note that $(pi(T))^{ri}(\mathbf{v}) = \mathbf{0}$ since $\mathbf{v} \in V_i$ where V_i is the null space of $(p_i(T))^{ri}$. Now we check $(pi(T))^{ri}(T\mathbf{v}) = T((p_i(T))^{ri}(\mathbf{v})) = T(\mathbf{0}) = \mathbf{0}$. Thus $T\mathbf{v} \in V_i$.

Theorem 3.2.6

Let the minimal polynomial of T be $\mu_T(x) = \prod_{i=1}^{k} (p_i(x))^{r_i}$ where $p_i(x)$ is irreducible for $i = 1, \ldots, k$, and V_i is the null space of linear transformations $(p_i(T))^{r_i}$. Then $V = V_1 \oplus \cdots \oplus V_k$, is a direct sum of T-invariant subspaces.

Proof: Let $q_i(x) = \mu_T(x)/p_i(x)^{r_i}$, so that $q_i(x)$ contains all but one of the irreducible factors of $\mu_T(x)$. Note that if $i \neq j$, then $\mu_T(x) \mid q_i(x)q_j(x)$. Since $q_1(x), \ldots, q_k(x)$ are relatively prime, $\gcd(\{q_i(x)\}_{i=1}^{k}) = 1$, there must exist polynomials $a_1(x), \ldots, a_k(x)$ such that $\sum_{i=1}^{k} a_i(x)q_i(x) = 1$.

Let $E_i = a_i(T)q_i(T)$. Then $\sum_{i=1}^{k} E_i = \sum_{i=1}^{k} a_i(T)q_i(T) = \mathbb{I}$. For $i \neq j$, $E_iE_j = \mu_T(T)q(T) = \mathbb{O}$, i.e., zero operator, for some polynomial $q(x)$. Thus

$$V = \mathbb{I}V = (E_1 + \cdots + E_k)V = E_1V + \cdots + E_kV.$$

We only need to show that $E_iV = V_i$ for all i. First show $E_iV \subseteq V_i$. Let $\mathbf{v} \in E_iV$. Then $\mathbf{v} = (a_i(T)q_i(T))(\mathbf{w})$, **for some** $\mathbf{w} \in V$, *so*

$$(P_i(T)^{ri})(\mathbf{v}) = (pi(T)^{r_i})(a_i(T)q_i(T))(\mathbf{w})$$

$$= a_i(T)(p_i(T)^{r_i} q_i(T))(\mathbf{w})$$
$$= a_i(T)\mu\mathrm{T}(T)\mathbf{w} = \mathbf{0}.$$

Thus, $E_iV \subseteq V_i$.

Now we show $V_i \subseteq E_iV$. *Let* $\mathbf{v} \in V$. *Then* $(p_i(T)^{ri})(\mathbf{v}) = \mathbf{0}$, and so for all $i, j \neq i$, $(q_j(T))(\mathbf{v}) = \mathbf{0}$, since $q_j(T)$ has $p_i(T)^{r_i}$ as a factor. Thus $E_j\mathbf{v} = \mathbf{0}$ for all $i, j \neq i$. But

$$\mathbb{I} = E1 + \cdots + E_k$$

so

$$\mathbf{v} = (E_1 + \cdots + E_k)\mathbf{v} = \sum_{i=1}^{k} E_i\mathbf{v} = \mathbf{0} + \cdots \mathbf{0} + E_i\mathbf{v} + 0 + \cdots + \mathbf{0} \in E_iV$$

Thus $V_i \subseteq E_iV$. Therefore, $V_i = E_iV$, and $V = V_1 + \cdots + V_k$.

Now, we will show that $V_i \cap V_j = \{0\}$. Without loss of generality, we will show the case for $i = 1, j \neq 1$. For this, we will show that $E_1 : V_1 \rightarrow V_1$ is a bijection. Let $\mathbf{v}_1 \in V_1$. Thus, there exists *a* $\mathbf{v} \in V$ such that $E_1\mathbf{v} = \mathbf{v}_1$. Since $\mathbf{v} = \mathbf{u}_1 + \mathbf{w}$ for some $\mathbf{u}_1 \in V_1$ and $\mathbf{w} \in V_2 + \cdots + V_k$. So $\mathbf{v}_1 = E_1\mathbf{v} = E_1(\mathbf{u}_1 + \mathbf{w}) = E_1\mathbf{u}_1 + E_1\mathbf{w} = E_1\mathbf{u}_1 + \mathbf{0} = E1\mathbf{u}_1$. Therefore, $E_1 : V_1 \rightarrow V_1$ is onto. Since dim V_1 is finite, E_1 is a bijection. Therefore, each E_i is a bijection.

Now, let $\mathbf{0} \neq \mathbf{v} \in V_i \subseteq V_j$ for $i \neq j$. $E_jE_i\mathbf{v} = \mathbf{0}$. However, E_i is a bijection, $\mathbf{u} = E_i\mathbf{v} \neq \mathbf{0}$. Since E_j is a bijection, $\mathbf{0} \neq E_j\mathbf{u} = E_jE_i\mathbf{v} = \mathbf{0}$, a contradiction. Thus, $V_i \cap Vj = \{\mathbf{0}\}$. Thus, $V = \oplus_{i=1}^{k} V_i$.

Theorem 3.2.7

Let $T : V \rightarrow V$ be a linear operator on a finite-dimensional vector space V over the field $\mathbb{F}$. Then T is diagonalizable if and only if μ_T is of the form

$$\mu_T(x) = (x - c_1) \ldots (x - c_k). \tag{3.1}$$

where $c_1, \ldots, c_k$ be the distinct eigenvalues of T.

Proof: If T *is diagonalizable, it has a basis formed by eigenvectors. Let* $\mathbf{b}$ be an eigenvector corresponding to an eigenvalue c_i. Then $T\mathbf{b} - c_i\mathbf{b} = \mathbf{0}$. So $x - c_i$ is a factor of the minimal polynomial, which can be computed just one linear factor for each of the distinct eigenvalue on the diagonal of T.

Conversely, let the minimal polynomial of T be $\mu_T(x) = \prod_{i=1}^{k}(x - c_i)$. Then V is a direct sum of the null spaces of $(T - c_i\mathbb{I})$, which shows that there exists a basis for V consisting of eigenvectors, and so T is diagonalizable.

Let A be a square $n \times n$ matrix with n linearly independent eigenvectors, $\mathbf{v}_1, \ldots, \mathbf{v}_n$. Then A can be factorized, or has an eigen-decomposition as

$$A = Q\Lambda Q^{-1}$$

where Q is the square $n \times n$ matrix whose i-th column is the eigenvector $\mathbf{v}_i$ of A and Λ is the diagonal matrix whose diagonal elements are the corresponding eigenvalues, i.e., $\Lambda_n = \lambda_i$. Note that only diagonalizable matrices can be factorized in this way. Essentially, A is a linear operator on $\mathbb{F}^n$ with respect to the standard basis, and Q is a change of bases matrix from basis formed by eigenvectors to standard basis.

3.3 Jordan Canonical Form

We recall that a subspace W of V is called T-invariant if $T(\mathbf{v}) \subseteq W$ for all $\mathbf{v} \in W$. A special case occurs with eigenvectors: the vector $\mathbf{v}$ is an eigenvector of T if and only if the 1-dimensional subspace it determines is a T-invariant subspace. Lets use these ideas in talking about diagonalization. Instead of saying that T can be diagonalized if and only if there is a basis for V consisting of eigenvectors of T, we can now say that T can be diagonalized if and only if V can be written as a direct sum of 1-dimensional T-invariant subspaces. It is this statement that can be generalized.

For a diagonalizable transformation, we have a very nice form. However it is not always true that a transformation is diagonalizable. For example, it may be that the roots of the characteristic polynomial are not in the field. Sometimes, even if the roots are in the field, they may not have multiplicities equal to the dimensions of the eigenspaces. So we look for a more general form. Instead of looking for a diagonal form, we look for a block diagonal form.

When the subspace W is invariant under the operator T, then T induces a linear operator T_W on the space W. The linear operator T_W is defined by $T_W(\alpha) = T(\alpha)$ for $\alpha \in W$. Note $T_W \neq T$ since the domain and T and T_w are different. When V is finite-dimensional, the invariance of W under T has a simple matrix interpretation. Suppose we choose an ordered basis $\mathcal{B}' = \{\alpha_1, \ldots, \alpha_n\}$ for V such that $\mathcal{B}' = \{\alpha_1, \ldots, \alpha_r\}$ is an ordered basis for $W (r = \dim W \leq n)$. Let $A = [T]_{\mathcal{B}}$ so that

$$T\alpha_j = \sum_{i=1}^{n} A_{ij}\alpha_i$$

Since W is invariant under T, the vector $T\alpha_j, \in W$ for all $j \leq r$. This means

$$T\alpha_j = \sum_{i=1}^{r} A_{ij}\alpha_i, \quad j \leq r.$$

This means that

$$A_{ij} = 0 \text{ if } j \leq r, \text{ and } i > r.$$

Thus A is a block matrix

$$A = \begin{bmatrix} B & C \\ 0 & D \end{bmatrix},$$

B is $r \times r$ matrix, C is $r \times (n - r)$ matrix and D is $(n - r) \times (n - r)$ matrix. B is the matrix of the induced operator T_W in the ordered basis $\mathcal{B}'$.

Definition 3.3.1

A Jordan normal form or Jordan canonical form of a linear operator J on a finite-dimensional vector space is an upper triangular matrix of block diagonal matrices J_i

$$J = \begin{bmatrix} J_1 & & \\ & \ddots & \\ & & J_P \end{bmatrix}$$

where each block J_i, namely Jordan block of A, is a square matrix of the form

$$J_i = \begin{bmatrix} \lambda_i & 1 & & \\ & \lambda_i & \ddots & \\ & & \ddots & 1 \\ & & & \lambda_i \end{bmatrix}, \text{ where } \lambda_i \text{ are the eigenvalues.}$$

In general, Jordan canonical form must be in block diagonal form, where each block has a fixed scalar on the main diagonal, and 1's or 0's on the superdiagonal. These blocks are called the **primary blocks**. The scalars for different primary blocks must be distinct. Each primary block must be made up of **secondary blocks** with a scalar on the main diagonal and only 1's on the superdiagonal (if a secondary block is of size 1×1, then it contains the scalar only). These blocks must be in order of decreasing size (moving down the main diagonal).

Definition 3.3.2

Let $W_1 \subseteq W_2$ be subspaces of V. Let $\mathcal{B}$ be a basis for W_1. Any set $\mathcal{B}'$ of vectors such that $\mathcal{B} \cup \mathcal{B}'$ is a basis for W_2 is called a complementary basis to W_1 in W_2.

Lemma 3.3.1: Let $T : V \to V$ be a linear transformation. Let $\{\mathbf{v}_1, \ldots, \mathbf{v}_k\}$ be a complementary basis to $N(T^m)$ in $N(T^{m+1})$, i.e., the null space of T^m in the null space of T^{m+1}. Then $\{T(\mathbf{v}_1), \ldots, T(\mathbf{v}_k)\}$ is part of a complementary basis to $N(T^{m-1})$ in $N(T^m)$.

Proof: Since $\mathbf{v}_i \in N(T^{m+1})$, *we have* $T^{m+1}(\mathbf{v}_i) = \mathbf{0}$, so $T^m(T(\mathbf{v}_i)) = \mathbf{0}$, and therefore $T(\mathbf{v}_i) \in N(T^m)$. Also, since $\mathbf{v}_i$ is in the complementary basis to $N(T^m)$ *in* $N(T^m{+}^1)$, $T^m(\mathbf{v}_i) \neq \mathbf{0}$.

We want to show that for each $i = 1, \ldots, k$, $T(\mathbf{v}_i) \in N(T^{m-1})$. Assume the opposite, i.e., $T^{m-1}(T(\mathbf{v}_i)) = \mathbf{0}$ for some i. But then $T^m(\mathbf{v}_i) = T^{m-1}(T(\mathbf{v}_i)) = \mathbf{0}$ which is a contradiction.

Now, we will show that $\{T(\mathbf{v}_1), \ldots, T(\mathbf{v}_k)\}$ are linearly independent. Suppose the

$$c_1T(\mathbf{v}_1) + \cdots + c_kT(\mathbf{v}_k) = \mathbf{0}, \text{ i.e., } c_1T(\text{v}1) + \cdots + c_kT(\mathbf{v}_k) \in N(T^{m-1}),$$

then

$$T^m(c_1\mathbf{v}_1 + \cdots + c_k\mathbf{v}_k) = T^{m-1}(c_1T(\mathbf{v}_1) + \cdots + c_kT(\mathbf{v}_k)) = \mathbf{0}$$

Hence $c_1\mathbf{v}_1 + \cdots + c_k\mathbf{v}_k \in N(T^m)$. Since $\{\mathbf{v}_1, \ldots, \mathbf{v}_k\}$ form a complementary basis to $N(T^m)$ in $N(T^m{+}^1)$, $c_s\mathbf{v}_1 + \ldots + c_k\mathbf{v}_k = \mathbf{0}$. But this implies that $c_i = 0$ for all i since $\{\mathbf{v}_1, \ldots, \mathbf{v}_k\}$ is a part of basis for $N(T^{m+1})$, thus linearly independent. This shows that $\{T(\mathbf{v}_1), \ldots, T(\mathbf{v}_k)\}$ are linearly independent, and also linearly independent of any vectors in $N(T^{m-1})$, so they are part of a complementary basis to $N(T^{m-1})$ in $N(T^m)$.

Theorem 3.3.1

(Jordan canonical form) If the characteristic polynomial of T is a product of linear factors, then it is possible to find a matrix representation for T which has Jordan canonical form.

In particular, suppose that $f(x) = (x - c_1)^{d_1} \ldots (x - c_k)^{d_k}$ is the characteristic polynomial of T and suppose that $\mu_T(x) = (x - c_1)^{r_1} \ldots (x - c_k)^{r_k}$ is the minimal polynomial of T. Then the matrix for T has primary blocks of size $d_i \times d_i$ corresponding to each eigenvalue c_i, and in each primary block the largest secondary block has size $r_i \times r_i$.

Proof: By the primary decomposition Theorem 3.2.6, we can obtain a block diagonal matrix, and then we can deal with each primary block separately. This reduces the proof to the case in which the characteristic polynomial is $(x - c)^d$ and the minimal polynomial is $(x - c)^r$ where $0 < r \leq d$.

Choose a complementary basis $\mathbf{v}_1, \ldots, \mathbf{v}_k$ to $N((T - c\mathbb{I})^{r-1})$ in $N((T - c\mathbb{I})^r)$. By the previous lemma, we have

$$(T - c\mathbb{I})(\mathbf{v}_1), \ldots, (T - c\mathbb{I})(\mathbf{v}_k)$$

are linearly independent.

Starting with $\{(T-c\mathbb{I})(\mathbf{v}_i)\}_{i=1}^{k}$, we find additional vectors $\mathbf{w}_1, \ldots, \mathbf{w}_t$ to complete a complementary basis to $N(T-c\mathbb{I})^{r-2}$ in $N(T-c\mathbb{I})^{r-i}$. Then continue this procedure until we have found a basis for $N((T-c\mathbb{I})^r)$ (note, this is the eigenspace corresponding to the eigenvalue c). Then consider the ordered basis:

$$\begin{aligned}
\mathcal{B} = {} & (T-c\mathbb{I})^{r-1}(\mathbf{v}_1), (T-c\mathbb{I})^{r-2}(\mathbf{v}_1), \ldots, \mathbf{v}_1, \\
& (T-c\mathbb{I})^{r-1}(\mathbf{v}_2), (T-c\mathbb{I})^{r-2}(\mathbf{v}_2), \ldots, \mathbf{v}_2, \\
& \ldots \\
& (T-c\mathbb{I})^{r-1}(\mathbf{v}_k), (T-c\mathbb{I})^{r-2}(\mathbf{v}_k), \ldots, \mathbf{v}_k, \\
& (T-c\mathbb{I})^{r-2}(\mathbf{w}_1), (T-c\mathbb{I})^{r-3}(\mathbf{w}_1), \ldots, \mathbf{w}_1, \\
& \ldots \\
& (T-c\mathbb{I})^{r-2}(\mathbf{w}_t), (T-c\mathbb{I})^{r-3}(\mathbf{w}_t), \ldots, \mathbf{w}_t, \\
& \ldots \\
& \mathbf{u}_1, \ldots, \mathbf{u}_m.
\end{aligned}$$

Note $(T-c\mathbb{I})^{r-1}(\mathbf{v}_1), (T-c\mathbb{I})^{r-1}(\mathbf{v}_2), \ldots, (T-c\mathbb{I})^{r-1}(\mathbf{v}_k)$ are eigenvectors for T corresponding to the eigenvalue c. But there may be more eigenvectors corresponding to the eigenvalue c, which we denoted by $\mathbf{u}_1, \ldots, \mathbf{u}_m$. Finally, we write $T = c\mathbb{I} + (T = c\mathbb{I})$. Then,

$$T((T-c\mathbb{I})^j\mathbf{v}_i) = c(T-c\mathbb{I}))^j\,(\mathbf{v}_i) + (T-c\mathbb{I}))^j+^1(^j\mathbf{v}_i).$$

Hence the matrix representation of T relative to the basis $\mathcal{B}$ has the Jordan block form.

The following corollary follows directly from the proof of the above theorem.

Corollary 3.3.1: If a square matrix A has Jordan canonical form J, then A is similar to its Jordan canonical form, i.e., $A = PJP^{-1}$ for some invertible matrix P. Note that the columns of P are the basis vectors we constructed, in the proof above.

Theorem 3.3.2
Let A and B be $n \times n$ matrices over the field of complex numbers. Then A and B are similar if and only if they have the same Jordan canonical form, up to the order of the eigenvalues.

Proof: The assumption that the matrices have complex entries guarantees that their characteristic polynomials can be factored completely, and so they can be put into Jordan canonical form.

If A and B are similar, then they represent the same linear transformation T. Therefore, A and B have the same characteristic polynomial, and the same minimal polynomial. Since the Jordan canonical form is constructed from the zeros of the characteristic polynomial of T, and the dimensions of the corresponding generalized eigenspaces, which are given by the minimal polynomial, therefore, the primary Jordan blocks are the same. However, we have no way to specify the order of the eigenvalues. Thus A and B have the same Jordan canonical form, except for the order of the eigenvalues.

Conversely, if A and B have the same Jordan canonical form J_A and J_B respectively, except for the order of the eigenvalues, then the matrices in Jordan canonical form are easily seen to be similar, i.e., $J_A \sim J_B$. By the previous corollary, $A \sim J_A \sim J_B \sim B$. Since similarity is an equivalence relation, it follows that A and B are similar.

Hence, in summary, suppose that T is a linear operator on V and that the characteristic polynomial for T can be factored as

$$f = (x - c_1)^{d_1} \cdots (x - c_k)^{dk}, \text{ where } c_i \text{ are distinct elements in } \mathbb{F},\ d_i \geq 1.$$

Then the minimal polynomial of T *is*

$$p = (x - c_1)^{r_1} \cdots (x - c_k)^{r_2}, \quad 1 \leq r_i < d_i$$

If W_i is the null space of $(T - c_i\mathbb{I})^{ri}$, then the primary decomposition says

$$V = W_1 \oplus \cdots \oplus W_k$$

and that the operator $T_i = T|W_i$ has minimal polynomial $(x - c_i)^{r_i}$. *Let* N_i be the linear operator on W_i defined by $N_i = T_i - c_i\mathbb{I}$. Then N_i has minimal polynomial x^{r_i}. On W_i, T acts like $N_i + c_i\mathbb{I}$. Suppose, we choose a basis for the subspace W_i such that the matrix of T_i in this ordered basis will be the direct sum of matrices

$$\begin{bmatrix} c & 1 & 0 & \cdots & 0 \\ 0 & c & 1 & \cdots & 0 \\ 0 & 0 & c & \cdots & 0 \\ \vdots & \vdots & & \ddots & 1 \\ 0 & 0 & 0 & \cdots & c \end{bmatrix}, \text{ with } c = c_i$$

The size of the matrices will decrease as one reads from left to write. A matrix of the form is called an elementary Jordan matrix with characteristic value c.

Let $A = \begin{bmatrix} A_1 & 0 & \cdots & 0 \\ 0 & A_2 & \cdots & 0 \\ \vdots & \vdots & & \vdots \\ 0 & 0 & \cdots & A_k \end{bmatrix}$ of matrices $A_1, \ldots, A_k$, where each A_i *is* of the form

$$A_i = \begin{bmatrix} J_1^{(i)} & 0 & \cdots & \\ 0 & J_2^{(i)} & \cdots & 0 \\ \vdots & \vdots & & \vdots \\ 0 & 0 & \cdots & J_{n_i}^{(i)} \end{bmatrix},$$

Where $J_j^{(i)}$ is an elementary Jordan matrix with characteristic value c_i. The sizes of the matrices $J_j^{(i)}$ decrease as j increases. An $n \times n$ matrix A that satisfies the conditions above is said to be in Jordan form.

If B is an $n \times n$ matrix over the field $\mathbb{F}$, and if the characteristic polynomial of B factors completely over $\mathbb{F}$, then B is similar over $\mathbb{F}$ to an $n \times n$ matrix A in Jordan form, and A is unique up to a rearrangement of the order of its characteristic values. We call A the Jordan form of B.

If A is a real matrix, its Jordan form can still be non-real since the eigenvalues may be complex numbers. We here give a real Jordan form for a matrix. Note that the complex roots always come as a pair of conjugate complex numbers. Let $\lambda_k = \alpha_k + i\beta_k$ and $\overline{\lambda}_k = \alpha_k - i\beta_k$ be a pair of complex conjugate eigenvalues. Let $\mathbf{v} = \mathbf{a} + i\mathbf{b}$ be an eigenvector of $\alpha_k + i\beta_k$, then

$$A(\mathbf{a} + i\mathbf{b}) = (\alpha_k + i\beta_k)(\mathbf{a} + i\mathbf{b}) = (\alpha_k\mathbf{a} - \beta_k\mathbf{b}) + i(\beta_k\mathbf{a} + \alpha_k\mathbf{b}).$$

Comparing the real part and imagery part of the two sides of this equation, $A\mathbf{a} = \alpha_k\mathbf{a} - \beta_k\mathbf{b}$ and $A\mathbf{b} = \beta_k a + \alpha_k\mathbf{b}$. We may represent the complex vector $\mathbf{v} = \mathbf{a} + i\mathbf{b}$ as a matrix $[\mathbf{a}\ \mathbf{b}]$ where the first column represents the real part of $\mathbf{v}$, and the second column the imaginary part of $\mathbf{v}$. Thus, each complex vector $\mathbf{v}$ can be represented by a pair of two real vectors $\mathbf{a}$ and $\mathbf{b}$. Writing this in matrix form, we have

$$A[a \quad b] = [a \quad b]\begin{bmatrix} \alpha_k & \beta_k \\ -\beta_k & \alpha_k \end{bmatrix}$$

For each pair of conjugate eigenvalues, we find a such pair of vectors. For the eigenvalue that has multiplicity greater than one, we can find the vectors using a process similar to the process used in Theorem 3.3.1. Note that all the vectors are real and the Jordan block for a complex eigenvalue $\alpha + i\beta$ with multiplicity m is

$$J_k = \begin{bmatrix} \alpha & \beta & 1 & 0 & & & & \\ -\beta & \alpha & 0 & 1 & & & & \\ & & \alpha & \beta & \ddots & & & \\ & & -\beta & \alpha & & & & \\ & & & & \ddots & & 1 & 0 \\ & & & & & & 0 & 1 \\ & & & & & & \alpha & \beta \\ & & & & & & -\beta & \alpha \end{bmatrix}_{2m\times 2m}$$

3.4 Trace

Definition 3.4.1

If $A \in M_{n\times n}(\mathbb{F})$ is an $n \times n$ matrix over $\mathbb{F}$,

$$A = \begin{pmatrix} a_{11} & a_{12} & \cdots & a_{1n} \\ a_{21} & a_{22} & \cdots & a_{2n} \\ \vdots & \vdots & \cdots & \vdots \\ a_{n1} & a_{n2} & \cdots & a_{nn} \end{pmatrix}$$

then the trace of A *is the* sum of the diagonal entries of A, that is, $\text{trace}(A) = a_{11} + a_{22} + \cdots + a_{nn}$.

We have the following properties of the trace.

Theorem 3.4.1

Let $A, B \in M_{n\times n}(\mathbb{F})$, and $c \in \mathbb{F}$. Then
$\text{trace}(cA) = c\text{trace}(A)$, $\text{trace}(A + B) = \text{trace}(A) + \text{trace}(B)$,
$\text{trace}(AB) = \text{trace}(BA)$.

Proof: Let $A = (a_{ij}$, and $B = (b_{ij})$. Then

$$\text{trace}(cA) = \text{trace}(ca_{ij}) \sum_{i=1}^{n} ca_{ii} = c\sum_{i=1}^{n} a_{ii} = c\text{trace}(A).$$

$$\text{trace}(A + B) = \text{trace}\ (a_{ij} + b_{ij}) = \sum_{i=1}^{n}(a_{ii} + b_{ii}) = \sum_{i=1}^{n} a_{ii} + \sum_{i=1}^{n} b_{ii}$$

$$= \text{trace}(A) + \text{trace}(B).$$

And

$$\text{trace}(AB) = \text{trace}\left(\left[\sum_{k=1}^{n} a_{ik}b_{kj}\right]\right) = \sum_{i=1}^{n}\sum_{k=1}^{n} a_{ik}b_{ki} = \sum_{k=1}^{n}\sum_{i=1}^{n} b_{ki}a_1k$$

$$= \left(\left[\sum_{i=1}^{n} b_{ki}a_{ij}\right]\right) = \text{trace}(BA)$$

Corollary 3.4.1: If A is similar to B, then trace(A) = trace(B).

Proof: By definition, *A is similar to B, then* $A = P^{-1}BP$, and

$$\text{trace}(A) = \text{trace}(P^{-1}BP) = \text{trace}(P^{-1}PB) = \text{trace}(B).$$

It is known that a square matrix A is always similar to its Jordan form, that is $A = PJP^{-1}$ for some invertible matrix P. Given an eigenvalue λ_i, its geometric multiplicity is $\dim(\ker(A - \lambda_i\mathbb{I}))$, and it is the number of Jordan blocks corresponding to λ_i. The sum of the sizes of all Jordan blocks corresponding to the eigenvalue λ_i is its algebraic multiplicity. A is diagonalizable if and only if, for every eigenvalue λ of A, its geometric and algebraic multiplicities coincide.

With these properties,

$$\text{trace}(A) = \text{trace}(PJP^{-1}) = \text{trace}(J) = \sum_{i=1}^{k} \lambda_i \ \det(A) = \det(J) = \prod_{i=1}^{k} \lambda_i$$

3.5 Determinants

Recall definition of a ring (Definition 1.1.1), we can see that $M_{n\times n}(\mathbb{F})$ is a non-commutative ring with identity over $\mathbb{F}$.

Definition 3.5.1

Let K be a commutative ring with identity, n a positive integer, and let D be a function which assigns to each $n \times n$ matrix A over K a scalar $D(A) \in K$. We say that D is n-linear if for each i, $1 \leq i \leq n$, D is a linear function of the ith row when the other $(n - 1)$ rows are held fixed. That is

$$D(\alpha_1,\ldots,c\alpha_i + \alpha_i',\ldots,\alpha_n)$$
$$= cD(\alpha_1,\ldots,\alpha_i,\ldots,\alpha_n) + D(\alpha_1,\ldots,\alpha_i',\cdots,\alpha_n), i = 1,\ldots,n, c \in K$$

Lemma 3.5.1: A linear combination of n-linear function is n-linear.

Definition 3.5.2

Let D be an n-linear function. We say D is alternating or alternate if A' is a matrix obtained from A by interchanging two rows of A, then $D(A') = -D(A)$.

Note if D is alternate, then $D(A) = 0$ whenever two rows of A are equal.

Definition 3.5.3

Let K be a commutative ring with identity, and let n be a positive integer. Suppose D is a function from $n \times n$ matrices over K into K. We say that D is determinant function if D is n-linear, alternating, and $D(\mathbb{I}) = 1$.

Recall, a permutation is a function that reorders the set of integers {1, ..., n}. The value in the i-th position after the reordering σ is denoted by σ_i. The set of all such permutations (also known as the symmetric group on n elements) is denoted by S_n. For each permutation σ, sgn(σ) denotes the signature or sign of σ, a value that is +1 whenever the reordering given by a can be achieved by successively interchanging two entries an even number of times, and −1 whenever it can be achieved by an odd number of such interchanges. Below, we give the formal definition.

Definition 3.5.4

A sequence $(k_1, \ldots, k_n)$ of positive integers not exceeding n, with the property that no two of the k_i are equal, is called a permutation of degree n. The permutation is called even/odd if the number of transpositions (i.e., the interchange of two elements) used is either even or odd. The sign or signature of a permutation is

$$\text{sign}(\sigma) = \begin{cases} 1 & \text{if } \sigma \text{ is even} \\ -1 & \text{if } \sigma \text{ is odd} \end{cases}$$

Under the operation of composition, the set of permutations of degree n is a group. This group is usually called the symmetric group of degree n.

Definition 3.5.5

Let K be a commutative ring with identity and let n be a positive integer. There is precisely one determinant function on the set of $n \times n$ matrices over K and it is the function det defined by

$$\det(A) = \sum_{\sigma \in S_n} \text{sgn}(\sigma) \prod_{i=1}^{n} a_{i\sigma_i},$$

where the sum is computed over all permutations a of the set {1, 2, ..., n}

Remark: Below are some properties of determinant.

1. $\det(\mathbb{I}_n) = 1$ where $\mathbb{I}_n$ is the $n \times n$ identity matrix.

2. $\det(A^T) = \det(A)$.

3. $\det(A^{-1}) = \dfrac{1}{\det(A)} = \det(A)^{-1}$ if A^{-1} exists.

4. For square matrices A, B of equal size,

$$\det(AB) = \det(A)\det(B).$$

5. $\det(cA) = c^n \det(A)$ for an $n \times n$ matrix and $c \in K$.

6. If A is a triangular matrix, i.e. $a_{ij} = 0$ for $i > j$ (or, $i < j$), then

$$\det(A) = a_{11}a_{22}\cdots a_{nm} = \prod_{i=1}^{n} a_{ii}$$

7. Consider an $n \times n$ matrix as n columns, the determinant is an n-linear function. Hence, the elementary column operation—replacing column C_i by $C_i + \alpha C_j$ where $\alpha \in K$, and all other columns are left unchanged, then the determinant of A is unchanged. (A similar result holds for rows.)

8. If in a matrix, any row or column is 0, then the determinant of that particular matrix is 0. (A similar result holds for rows.)

9. If some column can be expressed as a linear combination of the other columns, its determinant is 0. (A similar result holds for rows.)

10. Suppose that A is a matrix of size $n \times n$ over a field $\mathbb{F}$, then the following are equivalent.

— A is non-singular.
— A row-reduces to the identity matrix.
— The null space of A contains only the zero vector.
— The linear system $A\mathbf{x} = \mathbf{b}$ has a unique solution for every possible choice of $\mathbf{b}$.
— The columns of A are a linearly independent set.
— A is invertible.
— The column space of A is $\mathbb{F}^n$.
— The columns of A are a basis for $\mathbb{F}^n$
— The rank of A is n.
— The nullity of A is zero.
— $\det(A) \neq 0$.

Definition 3.5.6

The i,j cofactor of a cofactor matrix $C = (C_{ij})$, is a signed version of a minor M_{ij} defined by

$$C_{ij} = (-1)^{i+j} M_{ij}$$

where M_{ij} is the determinant of a submatrix of original matrix A deleting the i-th row and j-th column. We can use this in the computation of the determinant of a matrix A by expansion of the j-column

$$\det(A) = \sum_{i=1}^{n} a_{ij}C_{ij}$$

The $n \times n$ matrix adj$A$ is the transpose of the matrix of cofactors of A, is called the classical adjoint of A.

$$(\text{adj}A)_{ij} = C_{ji} = (-1)^{i+j} M_{ji}.$$

One can write down the inverse of an invertible matrix by computing its cofactors by using Cramer's rule.

$$C = \begin{bmatrix} C_{11} & C_{12} & \cdots & C_{1n} \\ C_{21} & C_{22} & \cdots & C_{2n} \\ \vdots & \vdots & \ddots & \vdots \\ C_{n1} & C_{n2} & \cdots & C_{nn} \end{bmatrix}$$

Then the inverse of A is the transpose of the cofactor matrix times the reciprocal of the determinant of A:

$$A^{-1} = \frac{1}{\det(A)} C^T$$

Hence, we have the following theorem:

Theorem 3.5.1

Let A be an $n \times n$ matrix over K where K is a commutative ring with identity. Then A is invertible over K if and only if det A is invertible in K. When A is invertible, the unique inverse for A is

$$A^{-1} = (\det A)^{-1} \text{ adj}A.$$

In particular, an $n \times n$ matrix over a field is invertible if and only if the determinant is different from zero.

3.6 Generalized Companion Matrix

Let $p(t)$ be a $m \times n$ matrix with polynomial entries,

$$p(t) = c_0 + c_1(t) + \cdots + c_{d-1}t^{d-1} + c_d t^d, \quad c_i \in \mathbb{F}$$

Use the concept companion matrix, we define $n \times n$ matrices A, B over the field $\mathbb{K}$.

$$A = \begin{bmatrix} 0 & 1 & \cdots & 0 & 0 \\ 0 & 0 & 1 & \cdots & 0 \\ \vdots & \vdots & \ddots & \vdots & \vdots \\ 0 & 0 & 0 & \cdots & 1 \\ c_0 & c_1 & \cdots & c_{d-2} & c_{d-1} \end{bmatrix}, c_i \in \mathbb{F}$$

$$B = \begin{bmatrix} 1 & \cdots & 0 & 0 \\ 0 & 1 & \cdots & 0 \\ \vdots & \ddots & \vdots & \vdots \\ 0 & 0 & \cdots & 0 \\ 0 & \cdots & 0 & -c_d \end{bmatrix}$$

Then it is straightforward to compute

$$p(t) = (-1)^{d-1} \det(A - tB).$$

The matrix pencil (A, B) is called the companion pencil of $p(t)$ and can be used to compute the roots of $p(t)$ as the roots of $p(t)$ coincide with the eigenvalues of the pencil (A,B).

We can generalize this concept to polynomial matrices.

Definition 3.6.1

Let $P(t)$ be a $m \times n$ matrix with polynomial entries,

$$P(t) = C_0 + C_1 t + \cdots + C_{d-}1t^{d-1} + Cd^{td}, C_i \in \mathbb{F}^{m \times n}$$

Define the generalized companion matrix $((d-1)m + n) \times dm$ matrices A, B:

$$A = \begin{bmatrix} 0 & \mathbb{I} & \cdots & 0 & 0 \\ 0 & 0 & \mathbb{I} & \cdots & 0 \\ \vdots & \vdots & \ddots & \vdots & \vdots \\ 0 & 0 & 0 & \cdots & \mathbb{I} \\ C_0^T & C_i^T & \cdots & C_{d-2}^T & C_{d-1}^T \end{bmatrix}, C_i^T \text{ is the transpose,}$$

$$B = \begin{bmatrix} \mathbb{I} & \cdots & 0 & 0 \\ 0 & \mathbb{I} & \cdots & 0 \\ \vdots & \ddots & \vdots & \vdots \\ 0 & 0 & \cdots & \mathbb{I} \\ 0 & \cdots & 0 & -C_d^T \end{bmatrix} \mathbb{I} = \mathbb{I}_{m \times m} \text{ the identity matrix.}$$

Proposition 3.6.1: For all $t \in \mathbb{F}$ and all $\mathbf{v} \in \mathbb{F}^m$,

$$P^T(t)\mathbf{v} = \mathbf{0} \quad \Leftrightarrow \quad (A - tB)\begin{bmatrix} \mathbf{v} \\ t\mathbf{v} \\ \vdots \\ t^{d-1}\mathbf{v} \end{bmatrix} = \mathbf{0}$$

Proof: The straightforward computation will give the above property.

Without loss of generality, assume $m \le n$. Since rank$P(t)$ = rank$P^T(t)$, we say that $t_0 \in \mathbb{F}$ drops rank$P(t)$ if and only if rank$P(t_0) < m$. Hence we have

Theorem 3.6.1
rank$P(t) < m \quad \Leftrightarrow \quad$ rank$(A - tB) < dm$.

Proof: Suppose there exists a t_0 that drops rank$P(t)$, *i.e.*, $P(t_0) < m$. Then there exists $\mathbf{v} \neq 0$ such that $P(t)\mathbf{v} = \mathbf{0}$. Hence, this is equivalent to say that $(A - tB)\mathbf{x} = \mathbf{0}$ has a non-trivial solution. Hence rank$(A - tB) < dm$. Similarly, we have the other direction.

3.7 Solution of Linear System of Equations

Review the system of equations consists of linear equations:

$$\begin{cases} a_{11}x_1 + a_{12}x_2 + & \cdots & +a_{1n}x_n = b_1 \\ a_{21}x_1 + a_{22}x_2 + & \cdots & +a_{2n}x_n = b_2 \\ & \vdots & \\ a_{m1}x_1 + a_{m2}x_2 + & \cdots & a_{mn}x_n = b_m \end{cases} \tag{3.2}$$

The system can be written in the form of matrix $A\mathbf{x} = \mathbf{b}$ where

$$A = \begin{pmatrix} a_{11} & \cdots & a_{1n} \\ \vdots & \ddots & \vdots \\ a_{m1} & \cdots & a_{mn} \end{pmatrix}, \quad \mathbf{x} = \begin{pmatrix} x_1 \\ \vdots \\ x_n \end{pmatrix}, \quad \mathbf{b} = \begin{pmatrix} b_1 \\ \vdots \\ b_n \end{pmatrix}$$

If A *is a square matrix and nonsingular, then the linear system* $A\mathbf{x} = \mathbf{b}$ **has a unique solution $\mathbf{x} = A^{-1}\mathbf{b}$ for every possible choice of b.**

Theorem 3.7.1

(Cramer's rule). Let A be a nonsingular $n \times n$ matrix, the $A\mathbf{x} = \mathbf{b}$ has a unique solution $\mathbf{x} = [c_1, \ldots, c_n]^T$ with

$$c_k = \frac{\begin{vmatrix} a_{11} & \cdots & b_1 & \cdots & a_{1n} \\ \cdots & \cdots & \cdots & \cdots & \cdots \\ a_{n1} & \cdots & b_n & \cdots & a_{nn} \end{vmatrix}}{\begin{vmatrix} a_{11} & \cdots & a_{k1} & \cdots & a_{1n} \\ \cdots & \cdots & \cdots & \cdots & \cdots \\ a_{n1} & \cdots & b_{nk} & \cdots & a_{nn} \end{vmatrix}}, k = 1, \cdots n$$

where the numerator is the determinant of the matrix A in which the k-th column is replaced by $\mathbf{b}$.

Proof: A is a non-singular, the unique solution of $A\mathbf{x} = \mathbf{b}$ is $\mathbf{x} = A^{-1}\mathbf{b}$. Let $[c_1, \ldots, c_n]^T$ be the solution. Then, substitute $b_k = a_{k1}c_1 + \cdots + a_{kn}c_n$, and we have

$$\begin{vmatrix} a_{11} & \cdots & a_{11}c_1 + \cdots + a_{1n}c_n & \cdots & a_{1n} \\ \cdots & \cdots & \cdots & \cdots & \cdots \\ a_{n1} & \cdots & a_{n1}c_1 + \cdots a_{nn}c_n & \cdots & a_{nn} \end{vmatrix}$$

$$= c_k \begin{vmatrix} a_{11} & \cdots & a_{1k} & \cdots & a_{1n} \\ \cdots & \cdots & \cdots & \cdots & \cdots \\ a_{n1} & \cdots & a_{nk} & \cdots & a_{nn} \end{vmatrix},$$

since determinant is n-linear and alternating.

Hence, the formula in this theorem gives another presentation of the solution.

Let $\mathbf{b} = \mathbf{0}$, then $A\mathbf{x} = \mathbf{0}$ is a linear system of homogenous equations. The solutions of $A\mathbf{x} = \mathbf{0}$ form a linear space, we leave this proof to the interested readers.

The solutions of $A\mathbf{x} = \mathbf{b}$ and the solutions of $A\mathbf{x} = \mathbf{0}$ are closely related. Let x_0 be a solution of $A\mathbf{x} = \mathbf{b}$ and x_h be a solution of $A\mathbf{x} = \mathbf{0}$, then $x_0 + x_h$ is a solutions of $A\mathbf{x} = \mathbf{b}$. Suppose $\mathbf{x}_1 \neq \mathbf{x}_2$ are solutions of $A\mathbf{x} = \mathbf{b}$, then $\mathbf{x}_1 - \mathbf{x}_2$ is a solution of $A\mathbf{x} = \mathbf{0}$.

Theorem 3.7.2

Let A be an $m \times n$ matrix and S be the solution space of the homogenous linear system $A\mathbf{x} = \mathbf{0}$, then

$$\dim S + \text{rankA} = n.$$

Proof: Let dim $S = s$ and $\mathbf{v}_1, \ldots, \mathbf{v}_s$ form a basis of S. We extend $\mathbf{v}_1, \ldots, \mathbf{v}_s$ to a basis $\mathbf{v}_1, \ldots, \mathbf{v}_s, \mathbf{v}_{s+1}, \ldots, \mathbf{v}_n$ of $\mathbb{F}^n$. A induces a linear transformation from $\mathbb{F}^n$ to $\mathbb{F}^n$, $\mathbf{v} \rightarrow A\mathbf{v}$, and the image space is spanned by $A\mathbf{v}_{s+1}, \ldots, A\mathbf{v}_n$, so rank$A$ = dim Span$\{A\mathbf{v}_{s+1}, \ldots, A\mathbf{v}_n\}$.

Notice that the dimension of Span$\{A\mathbf{v}_{s+1}, \ldots, A\mathbf{v}_n\}$ is $n - s$, otherwise, there exist $\alpha_{s+1}, \ldots, \alpha_n$, not zero simultaneously, such that

$$A(\alpha_s{+}1\mathbf{v}_{s+1} + \cdots + \alpha_n\mathbf{v}_n) = \mathbf{0}.$$

This means that $\alpha_s{+}1\mathbf{v}_s{+}_1 + \cdots + \alpha_n\mathbf{v}_n \in S$, which is impossible since $\mathbf{v}_1, \ldots, \mathbf{v}_s, \mathbf{v}_{s+1}, \ldots, \mathbf{v}_n$ is a basis for $\mathbb{F}^n$.

Therefore, rank$A$ = dimSpan$\{A\mathbf{v}_{s+1}, \ldots, A\mathbf{v}_n\}$ = $n - s$, equivalently, dim S + rank$A = n$.

3.8 Some Applications of Matrices

We give some examples to illustrate the applications of matrices in different areas.

3.8.1 Linear Programming

Linear programming is an optimization method that is applied to a mathematical model with linear constraints. One of the most basic properties of linear program is the characterization of the program. In [Tiw04], the author shows the termination is related to eigenvalues of the corresponding matrix. We here review the results as an application of the eigenvalues and Jordan canonical form.

Consider a linear program (algorithm) of the form

$$P1 : \texttt{while}\ (B\mathbf{x} > \mathbf{0})\ \{\mathbf{x} := A\mathbf{x}\}$$

where $B\mathbf{x} > \mathbf{0}$ represents a conjunction of linear inequalities over the state variables $\mathbf{x}$ and $\mathbf{x} : = A\mathbf{x}$ represents the linear assignments to each of the variables. The program **P1** will terminate on a state $\mathbf{x} = \mathbf{c}$ if a $B\mathbf{c} < \mathbf{0}$. The Program P1 is said to terminate if it terminates on all initial values in $\mathbb{R}$ for the variables in $\mathbf{x}$.

Proposition **3.8.1** [Tiw04] If the linear loop program $P1$, defined by matrices A and B is non-terminating, then there exists a real eigenvector $\mathbf{v}$ of A, corresponding to positive eigenvalue, such that $B\mathbf{v} \geq \mathbf{0}$.

EXAMPLE 3.8.1

Let the linear program be

$$\texttt{while}\ ([1\ -1]\ \mathbf{x} > \mathbf{0})\left\{\mathbf{x} := \begin{bmatrix} -1 & 1 \\ 0 & 1 \end{bmatrix}\mathbf{x}\right\}$$

The matrix A has two eigenvalues −1 and 1. The eigenvector associated to the positive eigenvalue 1 is $\begin{bmatrix} 1 \\ 2 \end{bmatrix}$. Then $[1\ -1]\mathbf{x} = [\mathbf{1}\ -1]\begin{bmatrix} 1 \\ 2 \end{bmatrix} = -1 \not> 0$ and this program is terminating.

In the iteration computation induced by $\mathbf{x} := A\mathbf{x}$, we need to compute A^k for the k-th iteratiom. Jordan canonical form can help to reduce the computation.

Let P be an invertible matrix such that $P^{-1}AP = J$ where $J = Diag(J_1, J_2, \ldots, J_K)$ is the real Jordan form of A and each Jordan block J_i is either of the two forms:

$$\begin{bmatrix} \lambda_i & 1 & & \\ & \lambda_i & \ddots & \\ & & \ddots & 1 \\ & & & \lambda_i \end{bmatrix}, \begin{bmatrix} D_i & \mathbb{I} & & \\ & D_i & \ddots & \\ & & \ddots & \mathbb{I} \\ & & & D_i \end{bmatrix},$$

where $D = \begin{bmatrix} \alpha & \beta \\ -\beta & \alpha \end{bmatrix}$ and $\mathbb{I}$ is the 2×2 identity matrix. Then $P1$ can be re-written to an equivalent form

$$P2 : \text{while}\ (By > \mathbf{0})\ \{y := Jy\}$$

where $\mathbf{y} = P^{-1}\mathbf{x}$.

Suppose the $J_i = \begin{bmatrix} \lambda_i & 1 & & \\ & \lambda_i & \ddots & \\ & & \ddots & 1 \\ & & & \lambda_i \end{bmatrix}$ is of size $m_i \times m_i$, then the k-the power of this block can be given explicitly as

$$J_i^k = \begin{bmatrix} \lambda_i^k & \binom{k}{1}\lambda_i^{k-1} & \binom{k}{2}\lambda_i^{k-2} & \cdots & \binom{k}{m_i-1}\lambda_i^{k-m_i+1} \\ 0 & \lambda_i^k & \binom{k}{1}\lambda_i^{k-1} & \cdots & \binom{k}{m_i-2}\lambda_i^{k-m_i+2} \\ \vdots & \vdots & \ddots & \ddots & \vdots \\ 0 & 0 & \cdots & \lambda_i^k & \binom{k}{1}\lambda_i^{k-1} \\ 0 & 0 & \cdots & 0 & \lambda_i^k \end{bmatrix}$$

Note that λ_i can be replaced by D_i if $J_i = \begin{bmatrix} D_i & \mathbb{I} & & \\ & D_i & \ddots & \\ & & \ddots & \mathbb{I} \\ & & & D_i \end{bmatrix}$

3.8.2 Geometric Transformation

In computer graphics and computer-aided design, the transformations, such as translations, rotations, and scaling, are basic operations for geometric objects. These transformations can be easily implemented by multiplication of certain matrices- Let (x, y, z) be a point in $\mathbb{R}^3$, its homogeneous form in $\mathbb{PR}^3$ is $(x, y, z, 1)$. Given a point in homogeneous form (x, y, z, w), $w \neq 0$, its affine form is $(x/w, y/w, z/w) \in \mathbb{R}^3$.

Translation

For a given point, a translation means to move the point to a new position This can be mathematically achieved by adding a vector to a point; for instance, move $(x, y, z, 1)$ to $(x + x_0, y + y_0, z + z_0, 1)$. The translation of an arbitrary point $(x, y, z, 1)$ by a vector (x_0, y_0, z_0) can be implemented by multiplying $(x, y, z, 1)^T$ by the following matrix

$$T(x_0, y_0, z_0) = \begin{bmatrix} 1 & 0 & 0 & x_0 \\ 0 & 1 & 0 & y_0 \\ 0 & 0 & 1 & z_0 \\ 0 & 0 & 0 & 1 \end{bmatrix}$$

and we have $T_{(x_0,y_0,z_0)}(x, y, z, 1)^T = (x + x_0, y + y_0, z + z_0, 1)^T$

Scaling

Scaling is used to change the length of a vector. A scaling in homogeneous coordinate is to transform $(x, y, z, 1)$ to $(s_x x, s_y y, s_z z, 1)$. A geometric object is enlarged if its scaled parameters s_x, s_y, s_z are greater than one.

A scaling transformation can be implemented by multiplying $(x, y, z, 1)^T$ by the following matrix

$$S(s_x, s_y, s_z) = \begin{bmatrix} s_x & 0 & 0 & 0 \\ 0 & s_y & 0 & 0 \\ 0 & 0 & s_z & 0 \\ 0 & 0 & 0 & 1 \end{bmatrix}$$

and we have $S_{(s_x, s_y, s_z)}(x, y, z, 1)^T = (s_x x + s_y y, s_z z, 1)^T$

Rotations

Another transformation of geometric objects is rotation with respect to a coordinate axis. A rotation about the x-axis by angle a counter- clockwise with right-hand rule is represented by a rotation matrix:

$$R_{x,\alpha} = \begin{bmatrix} 1 & 0 & 0 & 0 \\ 0 & \cos(\alpha) & -\sin(\alpha) & 0 \\ 0 & \sin(\alpha) & \cos(\alpha) & 0 \\ 0 & 0 & 0 & 1 \end{bmatrix}$$

and we have

$$R_{x,a} = \begin{bmatrix} x \\ y \\ z \\ 1 \end{bmatrix} = \begin{bmatrix} x \\ y\cos(\alpha) - z\sin(\alpha) \\ y\sin(\alpha) + z\cos(\alpha) \\ 1 \end{bmatrix}$$

The matrix $R_{x,\alpha}$ is called the basic (or elemental) rotation matrix about the x-axis. Similarly, the basic rotation matrices about y axis and z axis are the following:

$$R_{y,\beta} = \begin{bmatrix} \cos(\beta) & 0 & -\sin(\beta) & 0 \\ 0 & 1 & 0 & 0 \\ \sin(\beta) & 0 & \cos(\beta) & 0 \\ 0 & 0 & 0 & 1 \end{bmatrix}$$

$$R_{z,\gamma} = \begin{bmatrix} \cos(\gamma) & -\sin(\gamma) & 0 & 0 \\ \sin(\gamma) & \cos(\gamma) & 0 & 0 \\ 0 & 0 & 1 & 0 \\ 0 & 0 & 0 & 1 \end{bmatrix}$$

The above transformations can be easily combined. A rigid transformation of a vector space preserves distances between every pair of points. A rigid transformation is obtained by combining only translations and rotations.

3.8.3 Matrices in Graph Theory

A graph G consists of a set of vertices $V(G) = \{v_1, v_2, \ldots, v_m\}$ *and edges* $E(G) = e_1, e_2, \ldots, e_n$. We generally represent a finite graph by its adjacency matrix $A(G) = (a_{ij})_{m\times m}$ which is a square matrix with the entries 0 or 1, precisely,

$$a_{ij} = \begin{cases} 1 & v_i v_j \in E(G) \\ 0 & v_i v_j \not\in E(G) \end{cases}$$

For example, a given graph is shown in Fig. 3.1;, its adjacency matrix is

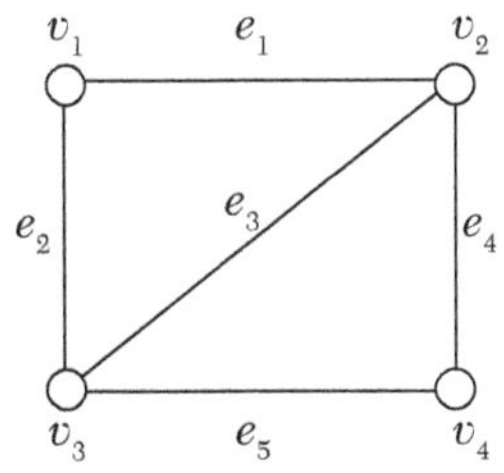

Figure 3.1 A simple graph

$$A(G) = \begin{bmatrix} 0 & 1 & 1 & 0 \\ 1 & 0 & 1 & 1 \\ 1 & 1 & 0 & 1 \\ 0 & 1 & 1 & 0 \end{bmatrix}$$

A commonly used property of the adjacency matrix is: the number of the paths from v_i to v_j with length n is exactly $a_{ij}^{(n)}$ where $A(G)^n = \left(a_{ij}^{(n)}\right)$..

For this example,

$$A(G)^2 = \begin{bmatrix} 2 & 1 & 1 & 2 \\ 1 & 3 & 2 & 1 \\ 1 & 2 & 3 & 1 \\ 2 & 1 & 1 & 2 \end{bmatrix}$$

$$A(G)^3 = \begin{bmatrix} 2 & 5 & 5 & 2 \\ 5 & 4 & 5 & 5 \\ 5 & 5 & 4 & 5 \\ 2 & 5 & 5 & 2 \end{bmatrix}$$

Check the paths from v_1 to v_2: there is one path with length one, one path path with length two and five pathes with length three. Note that we can use the same edge more than one time.

Another important matrix of the graph is Laplacian matrix $L(G)$. Let $D(G)$ be the degree matrix of a graph G, it is a diagonal matrix whose (i, i) entry is the degree of i-th vertex. Continue the example,

$$D(G) = \begin{bmatrix} 2 & 0 & 0 & 0 \\ 0 & 3 & 0 & 0 \\ 0 & 0 & 3 & 0 \\ 0 & 0 & 0 & 2 \end{bmatrix}$$

The Laplacian matrix of G is defined by $L(G) = D(G) - A(G)$ and

$$L(G) = \begin{bmatrix} 2 & -1 & -1 & 0 \\ -1 & 3 & -1 & -1 \\ -1 & -1 & 3 & -1 \\ 0 & -1 & -1 & 2 \end{bmatrix}$$

There are many properties of a Laplacian matrix such as

1. $L(G)$ is symmetric and positive-semidefinite;
2. The number of times 0 appears as an eigenvalue of $L(G)$ is the number of connected components in the graph;
3. The second smallest eigenvalue of $L(G)$ is the algebraic connectivity (or Fiedler value) of G.

For this simple example, the eigenvalues of $L(G)$ are 0, 2, 4 and 4. The number of connected components is obvious one and the algebraic connectivity of G is two according to the property of the Laplacian matrix.

3.8.4 Positional Relationship of Two Ellipsoids

Ellipsoids have a small number of geometric parameters and are excellent for approximating a wide class of convex objects in simulations of physical systems. Detecting a collision or overlap of two ellipsoids has practical applications in computer graphics, computer animation, virtual reality, robotics, CAD/CAM, etc. An algebraic equation of an ellipsoid can be written in a quadric form $X^TAX = 0$ with respect to a symmetric matrix A where

$X = (x, y, z, w)^T$. Based on discussion on the associated matrices, one can give an algebraic condition for the separation of ellipsoids in 3-dimensional Euclidean space, [WWK01].

Consider two ellipsoids $\mathcal{A} : X^T AX = 0$ and $\mathcal{B} : X^T BX = 0$ in homogeneous coordinates. Their characteristic polynomial is defined as

$$f(\lambda) = \det(\lambda A + B),$$

and $f(\lambda) = 0$ is called the characteristic equation. Assuming that the interior of $\mathcal{A}$ is defined by $X^T AX < 0$, and the interior of $\mathcal{B}$ is defined by $X^T BX < 0$ the algebraic conditions for the position relationships are:

The two ellipsoids are separated by a plane if and only if $f(\lambda) = 0$ has two distinct positive roots.

The two ellipsoids touch each other externally if and only if $f(\lambda) = 0$ has a positive double root.

Note that the characteristic equation $f(\lambda) = 0$ always has at least two negative roots. As soon as two distinct positive roots are detected, not necessary to compute the exact roots, one may conclude that the two ellipsoids are separated.

EXAMPLE 3.8.2

[WWK01] Consider the sphere $A : x^2 + y^2 + z^2 - 25 = 0$ and the ellipsoid $\mathcal{B} : (x - 9)^2/9 + y^2/4 + z^2/16 - 1 = 0$. Their associated symmetric matrices are

$$A = \begin{bmatrix} 1 & 0 & 0 & 0 \\ 0 & 1 & 0 & 0 \\ 0 & 0 & 1 & 0 \\ 0 & 0 & 0 & -25 \end{bmatrix}, B = \begin{bmatrix} \frac{1}{9} & 0 & 0 & -1 \\ - & \frac{1}{4} & 0 & 0 \\ 0 & 0 & \frac{1}{16} & 0 \\ -1 & - & 0 & 8 \end{bmatrix}$$

The four roots of the characteristic equation are −6.25, −1.5625, 0.60111 and 4.6211. Since there are two distinct positive roots, $\mathcal{A}$ and $\mathcal{B}$ are separated.

3.9 Exercises

1. Let $A = \begin{bmatrix} 1 & 2 & -1 \\ 3 & 5 & 4 \\ -3 & 1 & -2 \end{bmatrix}$. Find the sums of the principal minors of A of orders 1, 2, and 3, respectively.

2. Let $B = \begin{bmatrix} 1 & 1 & 1 \\ 2 & 3 & 4 \\ 5 & 8 & 9 \end{bmatrix}$. Find: det B, $adjB$, and find B^{-1} using adj$B$.

3. Let $A = \begin{bmatrix} a & b \\ c & d \end{bmatrix}$. Find: adj A. Show that adj(adj A) = A. When does $A = adjA$?

4. Find the characteristic polynomial of $A = \begin{bmatrix} 1 & 2 & 3 \\ 0 & 3 & 2 \\ 1 & 3 & 0 \end{bmatrix}$,

5. Consider the quadratic form $f(x, y) = 2x^2 - 4xy + 5y^2 = (x, y)A(x, y)^T$. Find matrix A. Can you diagonalize matrix A. Identify the change of variables so that $f(x, y)$ can be written as $AX^2 + BY^2$.

***Remark*:** Below is a solution provided by Andrew Crutcher via SAGE.

```
A = matrix(QQ,2,2,[2,-2,-2,5])
eigens = A.eigenvectors_right(); eigens
for eigen in eigens:
  print "Eigenvalue: ",eigen[0]
  print "Eigenvector: ",eigen[1][0]
  print "Multiplicity: ", eigen[2]
P = matrix.zero(QQ,A.nrows())
D = matrix.zero(QQ,A.nrows())
for coln in range(0,len(eigens)):
  v = (QQ"A.nrows()).zero_vector()
  v.set(coln,eigens[coln][0])
  D.set_column(coln,v)
  ev = eigens[coln][1][0]
  P.set_column(coln,ev)
P.inverse()*A*P
print "P"; P
print "D"; D
```

This would result in the following output.

```
[(6, [(1, -2)], 1), (1, [(1, 1/2)], 1)]
Eigenvalue: 6
Eigenvector: (1, -2)
Multiplicity: 1
```

```
Eigenvalue: 1
Eigenvector: (1, 1/2)
Multiplicity: 1
[6 0]
[0 1]
P
[ 1    1]
[ -2 1/2]
D
[6 0]
[0 1]
```

6. Find the minimal polynomial and characteristic polynomial of the matrix

$$A=\begin{bmatrix} 2 & 2 & -5 \\ 3 & 7 & -15 \\ 1 & 2 & -4 \end{bmatrix}.$$

Remark: Here we include the following solutions provided by Andrew Crutcher via SAGE.

```
A = matrix(QQ,3,3,[2,2,-5,3,7,-15,1,2,-4])
print "Eigenvalues:", A.eigenvalues()
print "Characteristic:", A.characteristic_polynomial()
print "Minimal:", A.minimal_polynomial()
This would result in the following output.
Eigenvalues: [3, 1, 1]
Characteristic: x^3 - 5*x^2 + 7*x - 3
Minimal: x^2 - 4*x + 3
```

7. Let $A=\begin{bmatrix} 2 & -4 \\ 2 & -6 \end{bmatrix}$. (a) Find all eigenvalues and corresponding eigenvectors. (b) Find matrices P and D is non-singular and $D=P^{-1}AP$ is diagonal.

Remark: We include a solution computed by SAGE provided by Andrew Crutcher.

```
A = matrix(QQ,2,2,[3,-4,2,-6])
eigens = A.eigenvectors_right(); eigens
for eigen in eigens:
  print "Eigenvalue: ",eigen[0]
  print "Eigenvector: ",eigen[1][0]
  print "Multiplicity: ", eigen[2]
P = matrix.zero(QQ,A.nrows())
D = matrix.zero(QQ,A.nrows())
```

```
for coln in range(0,len(eigens)):
  v = (QQ"A.nrows()).zero_vector()
  v.set(coln,eigens[coln][0])
  D.set_column(coln,v)
  ev = eigens[coln][1][0]
  P.set_column(coln,ev)
P.inverse()*A*P
print "P";
P print "D"; D
This would result in the following output.
[(2, [ (1, 1/4) ], 1), (-5, [ (1, 2) ], 1)]
Eigenvalue: 2
Eigenvector: (1, 1/4)
Multiplicity: 1
Eigenvalue: -5
Eigenvector: (1, 2)
Multiplicity: 1
[ 2  0]
[ 0 -5]
P
[  1 1]
[1/4 2]
D
[ 2  0]
[ 0 -5]
```

8. Let $A = \begin{bmatrix} 2 & 2 \\ 1 & 3 \end{bmatrix}$. (a) Find all eigenvalues and corresponding eigenvectors. (b) Find matrices P and D such that P is non-singular and $D = P^{-1}AP$ is diagonal. (c) Find A^6 *and* $f(A)$, where $f(t) = t^4 - 3t^3 - 6t^2 + 7t + 3$. (*d*) Find a matrix B such that $B^3 = A$ and B has real eigenvalues.

***Remark*:** Again, we include a solution by SAGE provided by Andrew Crutcher.

```
A = matrix(QQ,2,2,[2,2,1,3])
print "a)"
eigens = A.eigenvectors_right(); eigens
for eigen in eigens:
  print "Eigenvalue: ",eigen[0]
  print "Eigenvector: ",eigen[1][0]
  print "Multiplicity: ", eigen[2]
P = matrix.zero(QQ,A.nrows())
```

```
D = matrix.zero(QQ,A.nrows())
for coln in range(0,len(eigens)):
  v = (QQ"A.nrows()).zero_vector()
  v.set(coln,eigens[coln][0])
  D.set_column(coln,v)
  ev = eigens[coln][1][0]'
  P.set_column(coln,ev)
print "b)"
P.inverse()*A*P
print "P"; P
print "D"; D
print "c)"
print "A"6"; A"6
def f(x):
  return x"4-3*x"3-6*x"2+7*x+3
print "f(A)";f(A)
```

This would result in the following output.

```
a)
[(4, [(1, 1)], 1), (1, [(1, -1/2)], 1)]
Eigenvalue: 4
Eigenvector: (1, 1)
Multiplicity: 1
Eigenvalue: 1
Eigenvector: (1, -1/2)
Multiplicity: 1
b)
[4 0]
[0 1]
P
[ 1 1]
[ 1 -1/2]
D
[4 0]
[0 1]
c)
A"6
[1366 2730]
[1365 2731]
f(A)
[ 1 -2]
[-1  0]
```

For part *d*), let $C_{ij} = \sqrt{D_{ij}}, B = PCP$, and remember $A = PDP^{-1}$. Also note that if we take a diagonal matrix E and raise it to power ($E^3 = F$), realize that $F_{ij} = (E^3)_{ij} = (E_{ij})^3$ since when we do matrix multplication the only non-zero elements are on the diagonal and they will be only multiplied by themselves.

Note $B^3 = PCP^{-1}PCP^{-1}PCP^{-1} = PCCCP^{-1} = PC^3P^{-1} = PDP^{-1} = A$, so $B^3 = A$.

9. Let $A = \begin{bmatrix} 4 & -2 & 2 \\ 6 & -3 & 4 \\ 3 & -2 & 3 \end{bmatrix}$, and $B = \begin{bmatrix} 3 & -2 & 2 \\ 4 & -4 & 6 \\ 2 & -3 & 5 \end{bmatrix}$. Find characteristic polynomial of each matrix, and find the minimal polynomial of each matrix.

Remark:

```
A = matrix([[4, -2, 2],[6, -3, 4], [3, -2, 3]])
B = matrix([[3, -2, 2],[4, -4, 6], [2, -3, 5]])
Print "A: characteristic:", A.characteristic_polynomial()
Print "A: minimal poly: ", A.minimal_polynomial()
Print "B: characteristic:", B.characteristic_polynomial()
Print "B: minimal poly: ", B.minimal_polynomial()
```

This would result in the following output.

```
A:   characteristic: x^3 - 4*x^2 + 5*x - 2
A:   minimal poly: x^2 - 3*x + 2
B:   characteristic: x^3 - 4*x^2 + 5*x - 2
B:   minimal poly: x^3 - 4*x^2 + 5*x - 2
```

10. Find a matrix A whose minimal polynomial is $t^3 - 8t^2 + 5t + 7$.

Remark: The minimal polynomial of the companion matrix is the polynomial.

```
def f(x):
  return x^3-8*x^2+5*x+7
C = matrix([[0,1,0],[0,0,1],[-7,-5,8]]); C.transpose()
C.minimal_polynomial()
f(C)
```

This would result in the following output.

```
[0 0 -7]
[1 0 -5]
[0 1  8]
x^3 - 8*x^2 + 5*x + 7
[0 0 0]
```

```
[0 0 0]
[0 0 0]
```

11. Use the trace and the determinant of the matrix $A = \begin{bmatrix} 4 & 1 \\ -2 & 1 \end{bmatrix}$ to find the eigenvalues, and then deduce that it is diagonalizable. Find the eigenvalues of A^2 and A^{-1} if it exits.

12. Determine the Jordan normal form of $A = \begin{bmatrix} 4 & -2 & 2 \\ 6 & -3 & 4 \\ 3 & -2 & 3 \end{bmatrix}$, and $B = \begin{bmatrix} 3 & -2 & 2 \\ 4 & -4 & 6 \\ 2 & -3 & 5 \end{bmatrix}$.

Remark The following is a solution using Sage provided by Andrew Crutcher.

```
A = matrix([[4, -2, 2],[6, -3, 4], [3, -2, 3]])
B = matrix([[3, -2, 2],[4, -4, 6], [2, -3, 5]])
print "A Jordan form:"; A.jordan_form()
print "A Jordan form (no subdivides):";
   A.jordan_form(subdivide=false)
print "B Jordan form:"; B.jordan_form()
print "B Jordan form (no subdivides):";
   B.jordan_form(subdivide=false)
```

This would result in the following output.

A Jordan form:

```
[2|0|0]
[-+-+-]
[0|1|0]
[-+-+-]
[0|0|1]
A Jordan form (no subdivides):
[2 0 0]
[0 1 0]
[0 0 1]
B Jordan form:
[2|0 0]
[-+  ]
[0|1 1]
[0|0 1]
B Jordan form (no subdivides):
```

```
[2 0 0]
[0 1 1]
[0 0 1]
```

13. Find the eigenvectors and a basis of each eigenspace of the matrices over $\mathbb{R}$

$$A=\begin{bmatrix}4 & -5\\ 2 & -2\end{bmatrix}, B=\begin{bmatrix}1 & -1 & 1\\ 3 & 3 & -1\\ -6 & -2 & 4\end{bmatrix}, C=c,$$

$$D=A=\begin{bmatrix}5 & -1 & -1\\ 3 & 1 & -1\\ 6 & -2 & 0\end{bmatrix}, E=\begin{bmatrix}3 & 2 & -2 & 0\\ 2 & 3 & -2 & 0\\ 2 & 2 & -1 & 0\\ -2 & -4 & 6 & -1\end{bmatrix},$$

Are these matrices diagonalizable over $\mathbb{R}$. Find a Jordan normal form of them over $\mathbb{R}$, and a Jordan basis.

Remark: Solution provided by Andrew Crutcher via Sage is included below:

```
Eigenvalues[{{4, -5}, {2, -2}}]
Eigenvalues[{{1, -1, 1}, {3, 3, -1}, {-6, -2, 4}}]
Eigenvectors[{{1, -1, 1}, {3, 3, -1}, {-6, -2, 4}}]
```

This would result in the following output. Note I is $I=\sqrt{-1}$ for Wolfram

```
{1+1, 1-1}
{3 + 1 Sqrt[5], 3-1 Sqrt[5], 2}
{{-(1/6) I (2 I + Sqrt[5]), -(1/2),
1}, {1/6 I (-2 I + Sqrt[5]), -(1/2), 1}, {0, 1, 1}}
```

Since B has a real eigenvalue, this eigenvalue has a corresponding eigenvector which would be basis for the eigenspace of that eigenvalue $\mathbf{v}$ = [0, 1, 1]. Note since A, B have complex eigenvalues this means we cannot diagonalize them over $\mathbb{R}$ or find their Jordan form over $\mathbb{R}$. Going back to SAGE, we can compute the rest of the eigenvalues and eigenvectors.

```
C = matrix(QQ,[[1,1,0],[-2,4,0],[1,-1,2]])
D = matrix(QQ,[[5,-1,-1],[3,1,-1],[6,-2,0]])
E=matrix(QQ,[[3,2,-2,0],[2,3,-2,0],[2,2,-1,0],[-2,-4,6,-1]])

mats = [C,D,E]
```

```
for matlndex in range(0,len(mats)):
  print "Matrix",chr(ord('C')+matIndex)
  print "Eigenvalues: ", mats[matIndex].eigenvalues()
  for eigenInfo in mats[matIndex].eigenvectors_right():
    print "Eigenvalue: ",eigenInfo[0]
    print "Eigenvectors (basis of eigenspace for this
           eigenvalue):"
    print eigenInfo[1]
    #print eigen
print "Diagonalizable: ", mats[matIndex].is_diagonalizable()
print "Jordan Form"
mats[matIndex].jordan_form()
This would result in the following output.
Matrix C
Eigenvalues: [3, 2, 2]
Eigenvalue: 3
Eigenvectors (basis of eigenspace for this eigenvalue):
[
(1, 2, -1)
]
Eigenvalue: 2
Eigenvectors (basis of eigenspace for this eigenvalue):
[
(1, 1, 0),
(0, 0, 1)
]
Diagonalizable: True
Jordan Form
[3|0|0]
[-+-+-]
[0|2|0]
[-+-+-]
[0|0|2]
Matrix D
Eigenvalues: [2, 2, 2]
Eigenvalue: 2
Eigenvectors (basis of eigenspace for this eigenvalue):
[
(1, 0, 3),
(0, 1, -1)
]
Diagonalizable: False Jordan Form [2 1|0]
```

```
[0 2|0]
[+-]
[0 0|2]
Matrix E
Eigenvalues: [3, -1, 1, 1]
Eigenvalue: 3
Eigenvectors (basis of eigenspace for this eigenvalue):
[
(1, 1, 1, 0)
]
Eigenvalue: -1
Eigenvectors (basis of eigenspace for this eigenvalue):
[
(0, 0, 0, 1)
]
Eigenvalue: 1
Eigenvectors (basis of eigenspace for this eigenvalue):
[
(1, 0, 1, 2),
(0, 1, 1, 1)
]
Diagonalizable: True Jordan Form [ 3| 0| 0| 0]
[--+--+--+--]
[ 0|-1| 0| 0]
[--+--+--+--]
[ 0| 0| 1| 0]
[--+--+--+--]
[ 0| 0| 0| 1]
```

14. Find the eigenvectors and a basis of each eigenspace of the matrices over C

$$A = \begin{bmatrix} 4 & -5 \\ 2 & -2 \end{bmatrix}, B = \begin{bmatrix} 2 & 1 & 0 & 0 \\ 1 & 2 & 0 & 0 \\ 1 & 1 & 1 & 0 \\ 0 & -2 & 2 & -1 \end{bmatrix}$$

Are these matrices diagonalizable over $\mathbb{C}$? Find a Jordan normal form of them over $\mathbb{C}$, and a Jordan basis.

Remark: A solution provided by Andrew Cruther is included below via Sage.

```
A = matrix([[4, -5],[2,-2]])
B = matrix([[2,1,0,0],[1,2,0,0],[1,1,1,0],[0,-2,2,-1]])
```

```
mats = [A,B]
for matIndex in range(0,len(mats)):
  print "Matrix",chr(ord('AC')+matIndex)
  print "Eigenvalues: ", mats[matIndex].eigenvalues()
  for eigenInfo in mats[matIndex].eigenvectors_right():
    print "Eigenvalue: ",eigenInfo[0]
    print "Eigenvectors (basis of eigenspace for this
           eigenvalue):"
    print eigenInfo[1]
    print "Multiplicty: ", eigenInfo[2]
```

This would result in the following output.

```
Matrix A
Eigenvalues: [1 - 1*I, 1 + 1*I]
Eigenvalue: 1 - 1*I
Eigenvectors (basis of eigenspace for this eigenvalue):
[(1, 0.6000000000000000? + 0.2000000000000000?*I)]
Multiplicty: 1
Eigenvalue: 1 + 1*I
Eigenvectors (basis of eigenspace for this eigenvalue):
[(1, 0.6000000000000000? - 0.2000000000000000?*I)]
Multiplicty: 1
Matrix B
Eigenvalues: [3, -1, 1, 1]
Eigenvalue: 3
Eigenvectors (basis of eigenspace for this eigenvalue):
[
(1, 1, 1, 0)
]
Multiplicty: 1
Eigenvalue: -1
Eigenvectors (basis of eigenspace for this eigenvalue):
[
(0, 0, 0, 1)
]
Multiplicty: 1
Eigenvalue: 1
Eigenvectors (basis of eigenspace for this eigenvalue):
[
(1, -1, 0, 1),
(0, 0, 1, 1)
]
Multiplicty: 2
```

Since for both matrices, the multiplicty of each eigenvalue is the same as the dimension of the eigenspace for that eigenvalue, this would mean that each matrix is diagonalizable over $\mathbb{C}$ and the diagonalization of each matrix is also the Jordan normal form. J_A will denote the Jordan form of matrix A.

$$J_A = \begin{bmatrix} 1-i & 0 \\ 0 & 1+i \end{bmatrix}, J_B = \begin{bmatrix} 3 & 0 & 0 & 0 \\ 0 & -1 & 0 & 0 \\ 0 & 0 & 1 & 0 \\ 0 & 0 & 0 & 1 \end{bmatrix}$$

CHAPTER 4

ORTHOGONAL BASES

4.1 Inner Product Spaces

Definition 4.1.1

An inner product space is a vector space V along with a function $\langle , \rangle$ called an inner product:

$$\langle , \rangle : V \times V \rightarrow \mathbb{F}, \text{ where } \mathbb{F} \text{ is a field,}$$

which is

1. **positive definite**, i.e., $\langle \mathbf{u}, \mathbf{u} \rangle > 0$ with equality if and only if $\mathbf{u} = \mathbf{0}$,
2. **symmetric**, *i.e.*, $\langle \mathbf{u}, \mathbf{v} \rangle = \langle \mathbf{v}, \mathbf{u} \rangle$,
3. **bilinear**, *i.e.*, for any $a, b \in \mathbb{F}$,
   $\langle a\mathbf{u} + \mathbf{v}, \mathbf{w} \rangle = a\langle \mathbf{u}, \mathbf{w} \rangle) + \langle \mathbf{v}, \mathbf{w} \rangle$, and $\langle \mathbf{u}, b\mathbf{v} + \mathbf{w} \rangle = b\langle \mathbf{u}, \mathbf{v} \rangle + \langle \mathbf{u}, \mathbf{w} \rangle$.

Thus an inner product is a positive definite, symmetric bilinear form on the vector space V.

There are a few definitions that arise from the inner product definition. First, the **norm** of $\mathbf{x} \in V$ is defined as

$$|\mathbf{x}| = \sqrt{\langle \mathbf{x}, \mathbf{x} \rangle}$$

It is well defined by the non-negativity axiom of the Definition of the inner product space. The norm can be considered as the length of the vector $\mathbf{x}$. One can prove the Cauchy-Schwarz inequality, that is:

$$|\langle \mathbf{x}, \mathbf{y} \rangle| \leq |\mathbf{x}| \cdot |\mathbf{y}|, \forall \mathbf{x}, \mathbf{y} \in V,$$

with equality if and only if $\mathbf{x}$ and $\mathbf{y}$ are linearly dependent. An immediate consequence of the Cauchy-Schwarz inequality is that it justifies defining the angle between two non-zero vectors $\mathbf{x}$ and $\mathbf{y}$ in the case $\mathbb{F} = \mathbb{R}$ by the identity

$$\text{angle}(\mathbf{x}, \mathbf{y}) = \arccos\frac{\langle \mathbf{x}, \mathbf{y}\rangle}{|\mathbf{x}|\cdot|\mathbf{y}|}$$

We will say that non-zero vectors $\mathbf{x}, \mathbf{y} \in V$ are **orthogonal** if and only if their inner product is zero, i.e., $\langle \mathbf{x}, \mathbf{y} \rangle = 0$.

Definition 4.1.2

Let W be a subspace of the inner product space V and let $\{\mathbf{w}_1, \ldots, \mathbf{w}_n\}$ be a basis for W such that $(\mathbf{w}_i, \mathbf{w}_j) = 0$ if $i \neq j$, then this basis is called an orthogonal basis. Furthermore, if $(\mathbf{w}_i, \mathbf{w}_i) = 1$ then this basis is called an orthonormal basis.

Lemma 4.1.1 If $\mathbf{v}_1, \ldots, \mathbf{v}_k \in V$ are non-zero, pairwise orthogonal vectors, i.e., $\mathbf{v}_i \neq 0$ and $(\mathbf{v}_i, \mathbf{v}_j) = 0$ for all $i \neq j$, then they are linearly independent.

Proof: Suppose

$$c_1\mathbf{v}_1 + \cdots + c_k\,\mathbf{v}_k = \mathbf{0}.$$

Let us take the inner product of this equation with any $\mathbf{v}_i$. Using linearity of the inner product and orthogonality, we compute

$$0 = \langle (c_1\mathbf{v}_1 + \cdots + c_k\mathbf{v}_k), \mathbf{v}_i\rangle = c_1\langle \mathbf{v}_1, \mathbf{v}_i\rangle + \cdots + c_k\,\langle \mathbf{v}_k, \mathbf{v}_i\rangle = c_i\langle \mathbf{v}_i, \mathbf{v}_i\rangle = c_i|\mathbf{v}_i|^2.$$

Therefore, if $\mathbf{v}_i \neq 0$, then $c_i = 0$. Since this holds for all $i = 1, \ldots, k$, the linear independence of $\mathbf{v}_1, \ldots, \mathbf{v}_k$ follows.

As a direct consequence, any collection of non-zero orthogonal vectors forms a basis for its span.

Theorem 4.1.1
Suppose $\mathbf{v}_1, \ldots, \mathbf{v}_k \in V$ are non-zero, pairwise orthogonal vectors of an inner product space V. Then $\mathbf{v}_1, \ldots, \mathbf{v}_k$ form an orthogonal basis for their span $W = \text{Span}\{\mathbf{v}_1, \ldots, \mathbf{v}_k\} \subseteq V$, which is therefore a subspace of dimension $k = \dim W$. In particular, if $\dim V = k$, then $\mathbf{v}_1, \ldots, \mathbf{v}_k$ form a orthogonal basis for V.

There are a few advantages of orthogonal and orthonormal bases. For instance, one of the key issues one encounters is to express other vectors as linear combinations of the basis vectors that is, to find their coordinates in the prescribed basis. In general, this is not an easy task, since this requires solving a system of linear equations. Computing a set of solutions may require a considerable effort in case of high-dimensional situations. However, this problem can be eliminated if the basis is orthogonal, or, even better, orthonormal. This is the crucial insight underlying the efficiency in computational problems.

Theorem 4.1.2

Let $\mathbf{v}_1, \ldots, \mathbf{v}_n \in V$ be an orthonormal basis for an inner product space V. Then one can write any vector $\mathbf{v} \in V$ as a linear combination

$$\mathbf{v} = c_1\mathbf{v}_1 + \cdots + c_n\mathbf{v}_n$$

in which its coordinates

$$c_i = \langle \mathbf{v}, \mathbf{v}_i \rangle \quad i = 1, \ldots, n,$$

are explicitly given as inner products. Moreover, its norm

$$|\mathbf{v}| = \sqrt{\sum_{i=1}^{n} c_i^2} = \sqrt{\sum_{i=1}^{n} \langle \mathbf{v}, \mathbf{v}_i \rangle^2}$$

Proof: Let us compute the inner product of

$$\mathbf{v} = c_1\mathbf{v}_1 + \cdots + c_n\mathbf{v}_n$$

with one of the basis vectors. Using the orthonormality conditions

$$\langle \mathbf{v}_j, \mathbf{v}_i \rangle = \left\langle \begin{matrix} 0 & i \neq j \\ 1 & i = j \end{matrix} \right.$$

Hence

$$\langle \mathbf{v}, \mathbf{v}_i \rangle = c_i.$$

To prove the norm formula, we use the property

$$|v| = \sqrt{\langle \mathbf{v}, \mathbf{v} \rangle} = \sqrt{\sum_{j=1}^{n}\sum_{i=1}^{n} c_i c_j \langle \mathbf{v}_i, \mathbf{v}_j \rangle} = \sqrt{\sum_{i=1}^{n} c_i^2} = \sqrt{\sum_{i=1}^{n} \langle \mathbf{v}, \mathbf{v}_i \rangle^2}$$

Since moving from an orthogonal basis to its orthonormal version is elementary, simply by dividing each basis vector by its norm, it is enough to use an orthogonal basis.

4.2 Gram-Schmidt Process to Produce Orthonormal Basis

We understand that it is more convenient to use orthogonal or orthonormal bases, a natural question arises: how can we construct them?

A practical algorithm known as the Gram-Schmidt process is one of the premier algorithms of applied and computational linear algebra.

In mathematics, particularly linear algebra and numerical analysis, the Gram-Schmidt process is a method for orthonormalizing a set of vectors in an inner product space, most commonly the Euclidean space $\mathbb{R}^n$, equipped with the standard inner product. The Gram-Schmidt process takes a finite, linearly independent set $S = \{\mathbf{v}_1, \ldots, \mathbf{v}_k\}$ for $k \leq n$ and generates an orthogonal set $S' = \{\mathbf{u}_1, \ldots, \mathbf{u}_k\}$ that spans the same k-dimensional subspace of $\mathbb{R}^n$ as S.

We define the projection operator by the vector projection of $\mathbf{v}$ onto $\mathbf{u}$:

$$\mathrm{proj}_{\mathbf{u}}(\mathbf{v}) = \frac{\langle \mathbf{v}, \mathbf{u} \rangle}{\langle \mathbf{u}, \mathbf{u} \rangle}\mathbf{u}$$

This operator projects the vector $\mathbf{v}$ orthogonally onto the line spanned by the vector $\mathbf{u}$. If $\mathbf{u} = \mathbf{0}$, we define proj$\mathbf{u}(\mathbf{v}) = \mathbf{0}$, i.e., the projection map is the zero map, sending every vector onto the zero vector. The Gram-Schmidt process then works as follows:

$$\mathbf{u}_1 = \mathbf{v}_1, \qquad \mathbf{w}_1 = \frac{\mathbf{u}_1}{|\mathbf{u}_1|}$$

$$\mathbf{u}_2 = \mathbf{v}_2 - \mathrm{proj}_{\mathbf{u}_1}(\mathbf{v}_2), \qquad \mathbf{w}_2 = \frac{\mathbf{u}_2}{|\mathbf{u}_2|}$$

$$\mathbf{u}_3 = \mathbf{v}_3 - \mathrm{proj}_{\mathbf{u}_1}(\mathbf{v}_3) - \mathrm{proj}_{\mathbf{u}_1\mathbf{u}_2}(\mathbf{v}_3), \qquad \mathbf{w}_3 = \frac{\mathbf{u}_3}{|\mathbf{u}_3|}$$

$$\mathbf{u}_4 = \mathbf{v}_4 - \mathrm{proj}_{\mathbf{u}_1}(\mathbf{v}_4) - \mathrm{proj}\mathbf{u}_2(\mathbf{v}_4) - \mathrm{proj}_{\mathbf{u}_3}(\mathbf{v}_4), \qquad \mathbf{w}_4 = \frac{\mathbf{u}_4}{|\mathbf{u}_4|}$$

$$\vdots \qquad \vdots$$

$$u_k = v_k - \sum_{j=1}^{k-1} \mathrm{proj}_{\mathbf{u}_j}(\mathbf{v}_k) \qquad \mathbf{w}_k = \frac{\mathbf{u}_k}{|\mathbf{u}_k|}$$

The vectors $\mathbf{u}_1, \ldots, \mathbf{u}_k$ are orthogonal vectors, and the normalized vectors $\mathbf{w}_1, \ldots, \mathbf{w}_k$ form an orthonormal set. The calculation of the sequence $\mathbf{u}_1, \ldots, \mathbf{u}_k$ is known as the Gram-Schmidt orthogonalization, while the calculation of the sequence $\mathbf{w}_1, \ldots, \mathbf{w}_k$ is known as the Gram-Schmidt orthonormalization as the vectors are normalized. One can check that this set of vectors indeed contains pairwise orthogonal unit vectors. The general proof proceeds by mathematical induction.

EXAMPLE 4.2.1

Consider $\mathbb{R}^3$ with the dot product as the inner product space. Let P be the plane spanned by vectors $\mathbf{x}_1 = (1, 2, 2)$ and $\mathbf{x}_2 = (1, 0, 2)$. Find an orthonormal basis for P, and extend it to an orthonormal basis for $\mathbb{R}^3$.

Proof: We note that $\mathbf{x}_1$, $\mathbf{x}_2$ is a basis for the plane P, since the two vectors are linearly independent over $\mathbb{R}$, and we can extend it to a basis for $\mathbb{R}^3$ by adding one vector from the standard basis. For instance, vectors $\mathbf{x}_1$, $\mathbf{x}_2$, $\mathbf{x}_3 = (0, 0, 1)$ form a basis for $\mathbb{R}^3$ because the determinant of the matrix formed by $\mathbf{x}_1$, $\mathbf{x}_2$, $\mathbf{x}_3$ is non-zero.

Now, using the Gram-Schmidt process, we orthogonalize the basis $\mathbf{x}_i$ as the following:

$$\mathbf{v}_1 = \mathbf{x}_1 = (1, 2, 2),$$

$$\mathbf{v}_2 = \mathbf{x}_2 - \frac{\langle \mathbf{x}_2, \mathbf{v}_1 \rangle}{\langle \mathbf{v}_2, \mathbf{v}_1 \rangle}\mathbf{v}_1 = (-1,0,2) - \frac{3}{9}(1,2,2) = (-4/3, -2/3, 4/3)$$

$$\mathbf{v}_3 = \mathbf{x}_3 - \frac{\langle \mathbf{x}_3, \mathbf{v}_1 \rangle}{\langle \mathbf{v}_1, \mathbf{v}_1 \rangle}\mathbf{v}_1 - \frac{\langle \mathbf{x}_3, \mathbf{v}_2 \rangle}{\langle \mathbf{v}_2, \mathbf{v}_2 \rangle}\mathbf{v}_2$$

$$= (0,0,1) - \frac{2}{9}(1,2w,2) - \frac{4/3}{4}(-4/3, -2/3, 4/3) = (2/9, -2/9, 1/9)$$

So $\{\mathbf{v}_1, \mathbf{v}_2, \mathbf{v}_3\}$ is an orthogonal basis for $\mathbb{R}^3$ while $\{\mathbf{v}_1, \mathbf{v}_2\}$ is an orthogonal basis for P. We only need to normalize the vectors,

$$\mathbf{w}_1 = \frac{\mathbf{v}_1}{|\mathbf{v}_1|} = (1/3, 2/3, 2/3),$$

$$\mathbf{w}_2 = \frac{\mathbf{v}_2}{|\mathbf{v}_2|} = (-2/3, -1/3, 2/3),$$

$$\mathbf{w}_3 = \frac{\mathbf{v}_3}{|\mathbf{v}_3|} = (2/3, -2/3, 1/3),$$

Therefore, $\{\mathbf{w}_1, \mathbf{w}_2\}$ is an orthonormal basis for P, and $\{\mathbf{w}_1, \mathbf{w}_2, \mathbf{w}_3\}$ is an orthonormal basis for $\mathbb{R}^3$.

Remark 4.2.1: Let W be a subspace of an inner product finite-dimensional vector space V. Let $S = \{\mathbf{v}_1, \ldots, \mathbf{v}_k\}$ be a basis for W, we can extend S to a basis $B = \{\mathbf{v}_1, \ldots, \mathbf{v}_k, \mathbf{v}_k+_1, \ldots, \mathbf{v}_n\}$ for V. By the Gram-Schmidt process, we obtain an orthogonal basis $B' = \{\mathbf{u}_1, \ldots, \mathbf{u}_k, \mathbf{u}_{k+1}, \ldots, \mathbf{u}_n\}$ for V. Note that $S' = \{\mathbf{u}_1, \ldots, \mathbf{u}_k\}$ is an orthogonal basis for W. Moreover, for any $\mathbf{a} \in \text{Span}(\mathbf{u}_{k+1}, \ldots, \mathbf{u}_n)$, $\mathbf{a}$ is orthogonal to any $\mathbf{b} \in W$. In the next section, we will see that $\{\mathbf{u}_{k+1}, \ldots, \mathbf{u}_n\}$ form a basis for the orthogonal complement of W in V.

4.3 Orthogonal Complements and Projections

In this section, we first introduce a concept of an orthogonal complement.

Definition 4.3.1

Let W be a subspace of the inner product space V. The **orthogonal complement** of W is the set

$$W^{\perp} = \{\mathbf{v} \in V | (\mathbf{v}, \mathbf{w}) = 0, \forall \mathbf{w} \in W\}.$$

Furthermore, let the projection of $\mathbf{x}$ onto $\mathbf{y}$ be

$$\mathbf{p} = \text{proj}_{\mathbf{y}}(\mathbf{x}) = \frac{\langle \mathbf{x}, \mathbf{y} \rangle}{\langle \mathbf{y}, \mathbf{y} \rangle}$$

Theorem 4.3.1

Suppose that $\mathbf{p}$ is the orthogonal projection of $\mathbf{x}$ onto the space spanned by $\mathbf{y}$. Then $\mathbf{p}$ and $\mathbf{x} - \mathbf{p}$ are orthogonal.

Proof: Since two vectors are orthogonal if and only if their inner product is zero, we check

$$\langle \mathbf{p}, \mathbf{x} - \mathbf{p} \rangle = \langle \mathbf{p}, \mathbf{x} \rangle - \langle \mathbf{p}, \mathbf{p} \rangle = \frac{\langle \mathbf{x}, \mathbf{y} \rangle}{\langle \mathbf{y}, \mathbf{y} \rangle} \langle \mathbf{y}, \mathbf{x} \rangle - \left(\frac{\langle \mathbf{x}, \mathbf{y} \rangle}{\mathbf{y}, \mathbf{y}} \right)^2 \langle \mathbf{y}, \mathbf{y} \rangle = 0$$

so $\mathbf{p}$ and $\mathbf{x} - \mathbf{p}$ are orthogonal.

Definition 4.3.2

Let W be a finite dimension subspace of the inner product space V and let $\{\mathbf{w}_1, \ldots, \mathbf{w}_n\}$ be an orthogonal basis for W. If $\mathbf{v}$ is any vector in V then the **orthogonal projection** of $\mathbf{v}$ onto W is the vector:

$$\mathbf{p}W = \sum_{i=1}^{n} \frac{\langle \mathbf{v}, \mathbf{w}_i \rangle}{\langle \mathbf{w}_i, \mathbf{w}_i \rangle} \mathbf{w}_i$$

If $\{\mathbf{w}_1, \ldots, \mathbf{w}_n\}$ is an orthonormal basis for W, then the orthogonal projection of $\mathbf{v}$ onto W is the vector:

$$\mathbf{p}W = \sum_{i=1}^{n} \langle \mathbf{v}, \mathbf{w}_i \rangle \mathbf{w}_i$$

We can prove the following theorem.

Theorem 4.3.2

Let V be an n-dimensional inner product space and W a subspace of V. The following hold:

1. $W^{\perp}$ is a subspace of V;
2. $W \cap W^{\perp} = \{\mathbf{0}\}$;
3. $\dim W + \dim W^{\perp} = n$;
4. $W \oplus W^{\perp} = V$.

Proof: (1) First note that $\mathbf{0} \in W^{\perp}$. Now suppose that $\mathbf{w}_1, \mathbf{w}_2 \in W^{\perp}$. Then $\langle \mathbf{w}_i, \mathbf{v}\rangle = 0$ for all $\mathbf{v} \in W$ for $i = 1, 2$. Moreover, for some $a, b \in \mathbb{F}$, $\langle a\mathbf{w}_1 + b\mathbf{w}_2, \mathbf{v}\rangle = a\langle \mathbf{w}_1, \mathbf{v}\rangle + b\langle \mathbf{w}_2, \mathbf{v}\rangle = 0$, so $a\mathbf{w}_1 + c\mathbf{w}_2 \in W^{\perp}$. Thus $W^{\perp}$ is a subspace of V.

(2) Let $\mathbf{v} \in W \cap W^{\perp}$, then

$$\langle \mathbf{v}, \mathbf{w}\rangle = 0,\ \forall\ \mathbf{w} \in W \Rightarrow \langle \mathbf{v}, \mathbf{v}\rangle = 0 \Rightarrow \mathbf{v} = 0 \Rightarrow W \cap W^{\perp} = \{\mathbf{0}\}.$$

(3) From Remark 4.2.1, $V = W + W^{\perp}$. By the second claim, and the dimension of subspaces

$$\dim V = \dim W + \dim W^{\perp} - \dim W \cap W^{\perp} \Rightarrow n = \dim W + \dim W^{\perp}$$

(4) From Remark 4.2.1 and the Definition of $W^{\perp}$, for every vector $\mathbf{v} \in V$, there are vectors $\mathbf{w} \in W$ and $\mathbf{u} \in W^{\perp}$ such that $\mathbf{v} = \mathbf{w} + \mathbf{u}$. If $\mathbf{v} = \mathbf{w}' + \mathbf{u}'$, then $\mathbf{w} - \mathbf{w}' = \mathbf{u}' - \mathbf{u} = 0 \in W \cap W^{\perp}$. Thus, $\mathbf{w} = \mathbf{w}'$ and $\mathbf{u} = \mathbf{u}'$. By the Definition of direct sum, for every vector $\mathbf{v} \in V$, there are unique vectors $W \in W$ and $\mathbf{u} \in W^{\perp}$ such that $\mathbf{v} = \mathbf{w} + \mathbf{u}$.

Now we will consider the following problem: Given an inner product vector space V, a subspace W, and a vector $\mathbf{v} \in V$, find the set of vectors $\mathbf{w} \in W$ which are the closest to $\mathbf{v}$, that is the norm $|\mathbf{v} - \mathbf{w}|$ is the smallest, i.e.,

$$\forall \mathbf{u} \in W,\ |\mathbf{v} - \mathbf{w}| \leq |\mathbf{v} - \mathbf{u}|.$$

Theorem 4.3.3

Let V be a vector space with inner product $\langle , \rangle$. Let $W \subseteq V$ be a subspace and $\mathbf{v} \in \mathbf{V}$. If $(\mathbf{v} - \mathbf{w}) \perp W$, then $|\mathbf{v} - \mathbf{w}| < |\mathbf{v} - \mathbf{u}|$ for all $\mathbf{u} \in W$ and $|\mathbf{v} - \mathbf{w}| = |\mathbf{v} - \mathbf{u}|$ if and only if $\mathbf{w} = \mathbf{u}$. Thus $\mathbf{w} \in W$ is closest to $\mathbf{v}$.

Proof: First we note that $|\mathbf{v} - \mathbf{w}| \leq |\mathbf{v} - \mathbf{u}|$ if and only if $|\mathbf{v} - \mathbf{w}|^2 \leq |\mathbf{v} - \mathbf{u}|^2$. Note that the square of the norm, $|\cdot|2 = \langle , \rangle$, and calculate

$$|\mathbf{v} - \mathbf{u}|^2 = |(\mathbf{v} - \mathbf{w}) + (\mathbf{w} - \mathbf{u})|^2 = |\mathbf{v} - \mathbf{w}|^2 + |\mathbf{w} - \mathbf{u}|^2$$

since $\mathbf{v} - \mathbf{w} \in W^{\perp}$, $\mathbf{w} - \mathbf{u} \in W$, and $(\mathbf{v} - \mathbf{w}, \mathbf{w} - \mathbf{u}) = 0$. Thus,

$$|\mathbf{v} - \mathbf{u}|^2 > |\mathbf{v} - \mathbf{w}|^2 \text{ since } |\mathbf{w} - \mathbf{u}|^2 \geq 0.$$

The equality in the above expression holds if and only if $|\mathbf{w} - \mathbf{u}|^2 = 0$ if and only if $\mathbf{w} = \mathbf{u}$.

Theorem 4.3.4

Let V be a vector space with inner product $\langle , \rangle$. Let $W \subseteq V$ be a subspace and $\mathbf{v} \in V$. If $\mathbf{w} \in W$ is the closest to $\mathbf{v}$, then $\mathbf{v} - \mathbf{w} \perp W$.

Proof: Since $\mathbf{w} \in W$ is the closest to $\mathbf{v}$, we know that $|\mathbf{v} - \mathbf{w}| \leq |\mathbf{v} - \mathbf{u}|$ for all $\mathbf{u} \in W$. Therefore the function $f : \mathbb{R} \to \mathbb{R}$ such that

$$f(t) = |\mathbf{v} - \mathbf{w} + t\mathbf{x}|^2 \quad \forall \mathbf{x} \in W,$$

has a minimum value at $t = 0$. We have

$$\begin{aligned} f(t) &= \langle \mathbf{v} - \mathbf{w} + t\mathbf{x}, \mathbf{v} - \mathbf{w} + t\mathbf{x} \rangle \\ &= \langle \mathbf{v} - \mathbf{w}, \mathbf{v} - \mathbf{w} \rangle + 2\mathrm{t}\langle \mathbf{v} - \mathbf{w}, \mathbf{x} \rangle + t^2\langle \mathbf{x}, \mathbf{x} \rangle \\ &= |\mathbf{v} - \mathbf{w}|^2 + 2t\langle \mathbf{v} - \mathbf{w}, \mathbf{x} \rangle + t^2|\mathbf{x}|^2. \end{aligned}$$

Thus, $f'(t) = 2(\mathbf{v} - \mathbf{w}, \mathbf{x}) + 2t|\mathbf{x}|^2$ and since 0 is a critical number, $f'(0) = 2(\mathbf{v} - \mathbf{w}, \mathbf{x}) = 0$. As $\mathbf{x} \in W$ is arbitrary, it follows that $\mathbf{v} - \mathbf{w} \perp W$.

The idea behind the construction of the vector $\mathbf{w}$ such that $\mathbf{v} - \mathbf{w} \pm W$ is the Gram-Schmidt orthogonalization process.

Theorem 4.3.5

Let V be a vector space with inner product $\langle , \rangle$. Let $W \subseteq V$ be a subspace and assume that $\{\mathbf{v}_1, \ldots, \mathbf{v}_n\}$ is an orthogonal basis for W. For $\mathbf{v} \in V$ let

$$w = \sum_{i=1}^{n} \frac{\langle \mathbf{v}, \mathbf{v}_i \rangle}{|\mathbf{v}_i|^2} \mathbf{v}_i$$

Then $\mathbf{v} - \mathbf{w} \perp W$ (or equivalently, $\mathbf{w}$ is the vector in W closest to $\mathbf{v}$).

Proof:

$$\begin{aligned} \langle \mathbf{v} - \mathbf{w}, \mathbf{v}_j \rangle &= \langle \mathbf{v}, \mathbf{v}_j \rangle - \langle \mathbf{w}, \mathbf{v}_j \rangle \\ &= \langle \mathbf{v}, \mathbf{v}_j \rangle - \sum_{i=1}^{n} \frac{\langle \mathbf{v}, \mathbf{v}_j \rangle}{|\mathbf{v}_i|^2} \langle \mathbf{v}_i, \mathbf{v}_j \rangle \\ &= \langle \mathbf{v}, \mathbf{v}_j \rangle - \frac{\langle \mathbf{v}, \mathbf{v}_j \rangle}{|\mathbf{v}_j|^2} \langle \mathbf{v}_j, \mathbf{v}_j \rangle \\ &= 0 \end{aligned}$$

Hence $\mathbf{v} - \mathbf{w} \perp \mathbf{v}_j$. Since $\{\mathbf{v}_1, \ldots, \mathbf{v}_n\}$ is a basis for W, this implies that $\mathbf{v} - \mathbf{w} \perp W$.

One of the application problems related to orthogonal projection is the numerical approach to understand the least squares problem. When we try to fit one line on more than two points, we tend to face a problem that the linear equation $A\mathbf{x} = \mathbf{b}$ has no solution because there are more equations than the number of variables. In other words, the linear system is inconsistent and $\mathbf{b}$ is not in the column space of A. To handle this situation, we can regard $A\mathbf{x}$ as an approximation of $\mathbf{b}$, the optimal solution is to obtain $\mathbf{x}$ such that $|\mathbf{b} - A\mathbf{x}|$ is the smallest. Hence, instead of solving $A\mathbf{x} = \mathbf{b}$, our goal is to find a vector $\mathbf{x}$ such that $|\mathbf{b} - A\mathbf{x}|$ is the smallest. Therefore, if A is a $m \times n$ matrix and $\mathbf{b} \in \mathbb{R}^m$, a least squares solution of $A\mathbf{x} = \mathbf{b}$ is a vector $\mathbf{x}^* \in \mathbb{R}^n$ such that $|A\mathbf{x}^* - \mathbf{b}| \leq |A\mathbf{y} - \mathbf{b}|$ for all $\mathbf{y} \in \mathbb{R}^n$.

To do this, we recall that the column space C of A coincides with the image (= range) of A, since if $A_1, \ldots, A_n$ are the columns of A and $\mathbf{e}_1, \ldots, \mathbf{e}_n$ are the standard unit coordinate vectors in $\mathbb{R}^n$, then $A\mathbf{e}_i = A_i$. Moreover if we assume that $\text{rank}A = n$, then $\ker(A) = \{\mathbf{0}\}$.

Theorem 4.3.6

For any $m \times n$ matrix A,

$$(\text{im}A)^{\perp} = \ker(A^T).$$

Proof: Let $V = \text{im}A \in \mathbb{R}^m$, where the columns of A are denoted by $A_1, \ldots, A_n$, then

$$\begin{aligned} V^{\perp} &= \{\mathbf{w} \in \mathbb{R}^m \mid A_i \cdot \mathbf{w} = 0, i = 1, \ldots, n\} \\ &= \{\mathbf{w} \in \mathbb{R}^m \mid A_i^T \mathbf{w} = 0, \ ^{i} = 1, \ldots, n\} \\ &= \{\mathbf{w} \in \mathbb{R}^m \mid A^T \mathbf{w} = \mathbf{0}\} \\ &= \ker(A^T). \end{aligned}$$

Theorem 4.3.7

(The approximation theorem). The orthogonal projection $\mathbf{p}W(\mathbf{x})$ is closer to $\mathbf{x}$ than any other vector of W.

Proof: For any $\mathbf{y} \in W$,

$$\mathbf{x} - \mathbf{y} = (\mathbf{x} - \mathbf{p}W(\mathbf{x})) + (\mathbf{p}W(\mathbf{x}) - \mathbf{y})$$

and $(\mathbf{x} - \mathbf{p}W(\mathbf{x})) \pm W$, and $\mathbf{p}W(\mathbf{x}) - \mathbf{y} \in W$. The Pythagorean theorem (for general Euclidean spaces) now shows that

$$|\mathbf{x} - \mathbf{y}|^2 = |\mathbf{x} - \mathbf{p}W(\mathbf{x})|^2 + |\mathbf{p}W(\mathbf{x}) - \mathbf{y}|^2 \geq |\mathbf{x} - \mathbf{p}W(\mathbf{x})|^2$$

with equality if and only if $\mathbf{y} = \mathbf{p}W(\mathbf{x})$.

Theorem 4.3.8

Let A be an $m \times n$ matrix with rank n, and let P denote an orthogonal projection onto the image of A. Then for every $\mathbf{y} \in \mathbb{R}^m$, the equation $\mathbf{Ax} = P\mathbf{y}$ has a unique solution $\mathbf{a} \in \mathbb{R}^n$. Moreover, $\mathbf{a}$ is the best approximate solution to the equation $\mathbf{Ax} = \mathbf{y}$, in the sense that for any $\mathbf{x} \in \mathbb{R}^n$, $|\mathbf{Aa} - \mathbf{y}| \leq |\mathbf{Ax} - \mathbf{y}|$ with equality if and only if $\mathbf{x} = \mathbf{a}$.

Proof: By definition, the orthogonal projection $P\mathbf{y}$ belongs to the image of A. Therefore $A\mathbf{a} = P\mathbf{y}$ for some $\mathbf{a} \in \mathbb{R}^n$. Moreover, $\mathbf{a}$ is uniquely determined, since if $A\mathbf{x}_1 = A\mathbf{x}_2$, then $A(\mathbf{x}_1 - \mathbf{x}_2) = \mathbf{0}$ and $\mathbf{x}_1 - \mathbf{x}_2 \in \ker(A)$, and $\mathbf{x}_1 = \mathbf{x}_2$ since $\ker(A) = \{\mathbf{0}\}$. By the approximation theorem, we know that $|P\mathbf{y} - \mathbf{y}| \leq |A\mathbf{x} - \mathbf{y}|$ for every $\mathbf{x} \in \mathbb{R}^n$, with equality if and only if $A\mathbf{x} = P\mathbf{y}$. Substituting $A\mathbf{a} = P\mathbf{y}$, we obtain that $|A\mathbf{a} - \mathbf{y}| \leq |A\mathbf{x} - \mathbf{y}|$.

Theorem 4.3.9

If the null space of A is $\{\mathbf{0}\}$, then the solution of the normal system of equations $A^{\mathrm{T}}\ \mathbf{Ax} = \mathrm{AT}\mathbf{y}$ exists and equals the least squares solution of $\mathbf{Ax} = \mathbf{y}$.

Proof: Let A be a matrix of size $m \times n$. Note since $(A^TA)^T = A^TA$, so A^TA is a symmetric matrix. Thus, for an eigenvector $\mathbf{v}$ of A^TA, then $A^TA\mathbf{v} = \lambda\mathbf{v}$ for some λ, and

$$\lambda|\mathbf{v}|^2 = \mathbf{v}^T(\lambda\mathbf{v}) = \mathbf{v}^TA^TA\mathbf{v} = (A\mathbf{v}) \cdot (\mathbf{Av}) \geq 0.$$

Thus $\lambda \geq 0$. Suppose A^TA is singular, then there exists $\mathbf{v} \neq 0$ such that $A^TA\mathbf{v} = \mathbf{0}$, hence $0 = \mathbf{v}^TA^TA\mathbf{v} = (A\mathbf{v}) \cdot (A\mathbf{v})$. This implies that $A\mathbf{v} = \mathbf{0}$, i.e., the null space of A is not $\{\mathbf{0}\}$ contradicting to the condition. Thus, A^TA must be non-singular.

Then, the system $A^T\mathbf{Ax} = A^T\mathbf{y}$ has a unique solution of the form $\mathbf{a} = (A^TA)^{-1}A^T\mathbf{y}$. Thus $A^T(\mathbf{y} - A\mathbf{a}) = A^T\mathbf{y} - A^TA\mathbf{a} = \mathbf{0}$, hence $\mathbf{y} - A\mathbf{a} \in \ker A^T = (\mathrm{im}A)^{\perp}$. Thus, $\mathbf{y} - A\mathbf{a} = \mathbf{y} - P\mathbf{y}$ for an orthogonal projection P of $\mathbf{y}$ onto the im^A. Thus $\mathbf{A}\mathbf{a} = P\mathbf{y}$, and a is the least square solution for the system $A\mathbf{x} = \mathbf{y}$. Therefore, the solution of the normal system of equations $A^T\mathbf{Ax} = A^T\mathbf{y}$ exists and equals the least squares solution of $A\mathbf{x} = \mathbf{y}$.

From the proof of the above theorem, we have that the solution for the system $A^T\mathbf{Ax} = A^T\mathbf{y}$ is $\mathbf{a} = (A^TA)^{-1}A^T\mathbf{y}$, so $\mathbf{Aa} = A(A^TA)^{-1}A^T\mathbf{y}$. We have shown that a is the least square solution, then $A(A^TA)^{-1}A^T$ can be considered as the orthogonal projection matrix onto $\mathrm{im}A$. Thus, we conclude as the following.

Corollary 4.3.1: Let A be an $m \times n$ matrix, then the matrix of the orthogonal projection onto the image of A is $A(A^TA)^{-1}A^T$.

Now, as an application of orthogonal projection, let us see the least squares approximation problem: given n data $(x_1, y_1), \ldots, (x_n, y_n) \in \mathbb{R}^2$, we would like to find a line L of the form $y = mx + b$ which is the "closest fit" for the given data points, in the sense that the "least squares error" term

$$S(m,b) = \sum_{i=1}^{n} (mx_i + b - y_i)^2$$

is as small as possible. To find a formula for the "least squares regression line" L, we note that the system of n equations in the two unknowns m, b

$$m\mathbf{x} + \mathbf{b} = \mathbf{y},\ \mathbf{x} = (x_1, \ldots, x_n)^T,\ \mathbf{b} = (b, \ldots, b)^T,\ \mathbf{y} = (y_1, \ldots, y_n)^T$$

is over-determined. Assume that at least two of the the x_i's are distinct, then the matrix

$$A = \begin{bmatrix} x_1 & 1 \\ \vdots & \vdots \\ x_n & 1 \end{bmatrix},$$

has rank 2, and the system we are trying to solve can be written as

$$A\mathbf{v} = \mathbf{y}, \quad \mathbf{v} = \begin{bmatrix} m \\ b \end{bmatrix}$$

In general, $\mathbf{y}$ may not lie in the image of A, so this system typically has no solution. However, let P denote orthogonal projection onto the column space of A. Then Theorem 4.3.8 states that there is a unique solution $\mathbf{v} = (m_0, b_0)$ to the equation $A\mathbf{v} = P\mathbf{y}$, and this solution minimizes the quantity $|A\mathbf{v} - \mathbf{y}|^2$. Since $|A\mathbf{v} - \mathbf{y}|^2 = S(m, b)$, it follows that the line L is given by the equation $y = m_{0x} + b_0$.

Theorem 4.3.10

The least squares regression line $_L$ is given by the equation $y = m_{0x} + b_0$, where

$$m_0 = \frac{(\mathbf{x} - \overline{\mathbf{x}}) \cdot (\mathbf{y} - \overline{\mathbf{y}})}{(\mathbf{x} - \overline{\mathbf{x}}) \cdot (\mathbf{x} - \overline{\mathbf{x}})}, \quad b_0 = y - m_0\overline{x}$$

Where

$$\overline{x} = \frac{1}{n}\sum_{i=1}^{n} x_i, \quad \overline{y} = \frac{1}{n}\sum_{i=1}^{n} y_i, \quad \overline{x} = \begin{bmatrix} \overline{x} \\ \vdots \\ \overline{x} \end{bmatrix}, \quad \overline{y} = \begin{bmatrix} \overline{y} \\ \vdots \\ \overline{y} \end{bmatrix}$$

Proof: Theorem 4.3.9 states that the least squares solution $\mathbf{v}$ satisfies the equation $A^TA\mathbf{v} = A^T\mathbf{y}$ where $A = \begin{bmatrix} x_1 & 1 \\ \vdots & \vdots \\ x_n & 1 \end{bmatrix}$ and $\mathbf{v} = \begin{bmatrix} m_0 \\ b_0 \end{bmatrix}$. Compute

$$A^TA = \begin{bmatrix} \mathbf{x}\cdot\mathbf{x} & n\overline{x} \\ n\overline{x} & n \end{bmatrix}; \quad A^T\mathbf{y} = \begin{bmatrix} \mathbf{x}\cdot\mathbf{y} \\ n\overline{y} \end{bmatrix}, \text{ and solve } A^TA\mathbf{v} = A^T\mathbf{y}$$

$$\mathbf{v} = (A^TA)^{-1}A^T\mathbf{y} = \frac{1}{n\mathbf{x}\cdot\mathbf{x} - n^2\overline{x}^2}\begin{bmatrix} n & -n\overline{x} \\ -n\overline{x} & \mathbf{x}\cdot\mathbf{x} \end{bmatrix}\begin{bmatrix} \mathbf{x}\cdot\mathbf{y} \\ n\overline{y} \end{bmatrix} = \frac{\begin{bmatrix} \mathbf{x}\cdot\mathbf{y} - n\overline{x}\,\overline{y} \\ -\overline{x}\mathbf{x}\cdot\mathbf{y} + \mathbf{x}\cdot\mathbf{x}\overline{y} \end{bmatrix}}{\mathbf{x}\cdot\mathbf{x} - n\overline{x}^2}$$

Therefore

$$\mathbf{v} = \begin{bmatrix} m_0 \\ b_0 \end{bmatrix}, \quad m_0 = \frac{\mathbf{x}\cdot\mathbf{y} - n\overline{x}\,\overline{y}}{\mathbf{x}\cdot\mathbf{x} - n\overline{x}^2}, \quad b_0 = \frac{-x\mathbf{x}\cdot\mathbf{y} + \mathbf{x}\cdot\mathbf{x}\overline{y}}{n\mathbf{x}\cdot\mathbf{x} - n\overline{x}^2}$$

Now, note that $\mathbf{x}\cdot\overline{\mathbf{x}} = \overline{\mathbf{x}}\cdot\overline{\mathbf{x}} = n\overline{x}^2$ and $\mathbf{x}\cdot\overline{\mathbf{y}} = \overline{\mathbf{x}}\cdot\mathbf{y} = \overline{\mathbf{x}}\cdot\overline{\mathbf{y}} = n\overline{x}\,\overline{y}$. Thus

$$\mathbf{x}\cdot\mathbf{x} - n\overline{x}^2 = (\mathbf{x}-\overline{\mathbf{x}})\cdot(\mathbf{x}-\overline{\mathbf{x}}), \; \mathbf{x}\cdot\mathbf{y} - n\overline{x}\overline{y} = (\mathbf{x}-\overline{\mathbf{x}})(\mathbf{y}-\overline{\mathbf{y}}).$$

From the equations above it follows that

$$m_0 = \frac{\mathbf{x}\cdot\mathbf{y} - n\overline{x}\,\overline{y}}{\mathbf{x}\cdot\mathbf{x} - n\overline{x}^2} = \frac{(\mathbf{x}-\overline{\mathbf{x}})\cdot(\mathbf{y}-\overline{\mathbf{y}})}{(\mathbf{x}-\overline{\mathbf{x}})\cdot(\mathbf{x}-\overline{\mathbf{x}})}$$

Therefore the solution is

$$\mathbf{v} = \begin{bmatrix} m_0 \\ b_0 \end{bmatrix}, \quad \text{where } m_0 = \frac{(\mathbf{x}-\overline{\mathbf{x}})\cdot(\mathbf{y}-\overline{\mathbf{y}})}{(\mathbf{x}-\overline{\mathbf{x}})-(\mathbf{x}-\overline{\mathbf{x}})}, \quad b_0 = \overline{y} - m_0\overline{x}.$$

EXAMPLE 4.3.1

Suppose the three data points are (1, 2), (2, 5), (3, 7). Then

$$\overline{x} = 2, \quad \overline{y} = 14/3, \quad \mathbf{x} = (1,\ 2,\ 3)^T,\ \mathbf{y} = (2,\ 5,\ 7)^T$$
$$\mathbf{x} - \overline{\mathbf{x}} = (-1,\ 0,\ 1)^T\ \mathbf{y} - \overline{\mathbf{y}} = (-8/3, 1/3,\ 7/3)^T.$$

$$m_0 = \frac{(\mathbf{x}-\overline{\mathbf{x}})\cdot(\mathbf{y}-\overline{\mathbf{y}})}{(\mathbf{x}-\overline{\mathbf{x}})-(\mathbf{x}-\overline{\mathbf{x}})} = 5/2, \quad b_0 = \overline{y} - m_0\overline{x} = 14/3 - (5/2)(2) = -1/3.$$

Therefore, the least squares regression line is given by the equation

$$y = \frac{5}{2}x - \frac{1}{3}$$

4.4 Orthogonal Projections and Reflections

Matrices whose columns form an orthonormal basis of $\mathbb{R}^n$ relative to the standard Euclidean dot product play a distinguished role, they are called orthogonal matrices. Such matrices appear in a wide range of applications in geometry, physics, mechanics, and special functions. In particular, calculations involving spatial rotations are described in terms of orthogonal matrices, and orthogonal matrices are an essential ingredient in the QR algorithm for computing eigenvalues of matrices. In this section, we will study the orthogonal matrices.

Definition 4.4.1

A square matrix Q is called an **orthogonal matrix** if it satisfies

$$Q^TQ = \mathrm{I}$$

The orthogonality condition implies that one can easily invert an orthogonal matrix:

$$Q^{-1} = Q^T.$$

In fact, a matrix is orthogonal if and only if its inverse is equal to its transpose.

Theorem 4.4.1

A matrix Q is orthogonal if and only if its columns form an orthonormal basis with respect to the Euclidean dot product on $\mathbb{R}^n$.

Proof: Let $\mathbf{u}_1, \ldots, \mathbf{u}_n$ be the columns of Q. Then $\mathbf{u}_1^T, \ldots, \mathbf{u}_n^T$ are the rows of the transposed matrix Q^T. The (i, j)-th entry of the product Q^TQ is given as the product of the i-th row of Q^T times the j-th column of Q. Thus, the orthogonality requirement implies

$$\mathbf{u}_i \cdot \mathbf{u}_j = \mathbf{u}_i^T \cdot \mathbf{u}_j = \begin{cases} 0 & i \neq j \\ 1 & i = k \end{cases}$$

which are precisely the conditions for $\mathbf{u}_1, \ldots, \mathbf{u}_n$ to form an orthonormal basis.

Corollary 4.4.1: An orthogonal matrix has determinant $\det Q = \pm 1$.

Proof: We observe that

$$\det(Q^TQ) = \det(Q^T)\det(Q) = (\det(Q))2 = 1 = \det(\mathbb{I}) \Rightarrow \det(Q) = \pm 1.$$

Corollary 4.4.2: The product of two orthogonal matrices is also orthogonal.

Proof: Let Q_1, Q_2 be two orthogonal matrices, then

$$(Q_1Q_2)^T(Q_1Q_2) = Q_2^TQ_1^TQ_1Q_2 = Q_2^T(Q_1^TQ_1)Q_2 = Q_2^TQ_2 = \mathbb{I}$$

and so the product matrix is also orthogonal.

This property says that the set of all orthogonal matrices forms a group. The orthogonal group lies at the foundation of Euclidean geometry, as well as rigid body rotation, computer graphics and animation, and many other areas.

4.4.1 Orthogonal Projection Matrix

Let $\mathbf{a}_1, \ldots, \mathbf{a}_k$ linearly independent vectors in $\mathbb{R}^n$ and $W = \text{Span}(\mathbf{a}_1, \ldots, \mathbf{a}_k)$. We will study how to construct projections and reflections matrices, and use these matrices to find an orthogonal projection of a given vector onto W, and reflection of a given vector $\mathbf{x} \in \mathbb{R}^n$ through W.

To do so, we will note that the orthogonal projection of $\mathbf{x}$ onto W is in fact another vector $\mathbf{x}' \in W$ such that $(\mathbf{x} - \mathbf{x}')^{\perp}\ \mathbf{y}$ for all $\mathbf{y} \in W$. Since $W = \text{Span}(\mathbf{a}1, \ldots, \mathbf{a}_k)$, we must have

$$\mathbf{a}_i \cdot (\mathbf{x} - \mathbf{x}') = 0,\ \forall_i = 1, 2, \ldots, {}_k.$$

Now let A be an $n \times k$ matrix, $k \le n$, with columns $\mathbf{a}_1, \ldots, \mathbf{a}_k$ that are linearly independent vectors in $\mathbb{R}^n$. Then the above relation means

$$A^T(\mathbf{x} - \mathbf{x}') = \mathbf{0}, \Rightarrow A^T\mathbf{x} = A^T\mathbf{x}'$$

since $\mathbf{x}' \in W$, there exist $C = [c_1, \ldots, c_k]^T$ such that $\mathbf{x}' = A[c_1, \ldots, c_k]^T = AC$. Therefore,

$$A^T\mathbf{x} = A^T\mathbf{x}' = A^TAC.$$

Thus, by the proof of Theorem 4.3.9, we know that if the A has linearly independent columns, then A^TA is invertible. By the relation $A^T\mathbf{x} = (A^TA)^C$, we obtain $C = (A^TA)^{-1}(A^T\mathbf{x})$, and

$$\mathbf{x}' = AC = A(A^TA)^{-1}\,(A^T\mathbf{x}) = (A(A^TA)^{-1}A^T)\mathbf{x}.$$

The matrix

$$Q = A(A^TA)^{-1}A^T$$

is called the orthogonal projection matrix onto the subspace W. We check that Q has the following two properties:

$$Q^T = (A(A^TA)^{-1}A^T)^T = A(A^TA)^{-1}A^T = Q,$$

$$Q^2 = (A(A^TA)^{-1}A^T)(A(A^TA)^{-1}A^T) = A(A^TA)^{-1}A^T = Q.$$

EXAMPLE 4.4.1

Compute the projection matrix Q for the 2-dimensional subspace W of $\mathbb{R}^4$ spanned by the vectors $(1, 1, 0, 2)$ and $(-1, 0, 0, 1)$. What is the orthogonal projection of the vector $\mathbf{x} = (0, 2, 5, -1)$ onto W?

To answer the question, we let $A = \begin{bmatrix} 1 & -1 \\ 1 & 0 \\ 0 & 0 \\ 2 & 1 \end{bmatrix}$. Then

$$Q = A(A^T A)^{-1} A^T = \begin{bmatrix} 1 & -1 \\ 1 & 0 \\ 0 & 0 \\ 2 & 1 \end{bmatrix} \begin{bmatrix} 6 & 1 \\ 1 & 2 \end{bmatrix}^{-1} \begin{bmatrix} 1 & 1 & 0 & 2 \\ -1 & 0 & 0 & 1 \end{bmatrix}$$

$$= \begin{bmatrix} 10/11 & 3/11 & 0 & -1/11 \\ 3/11 & 2/11 & 0 & 3/11 \\ 0 & 0 & 0 & 0 \\ -1/11 & 3/11 & 0 & 10/11 \end{bmatrix}$$

Hence, the image of orthogonal projection is $Q\mathbf{x} = (7/11, 1/11, 0, -4/11)$T

4.4.2 Reflection Matrix

Now, let us use the result of orthogonal projection matrix to analyze what is the reflection of $\mathbf{x}$ across W, where W is a (hyper-)plane through the origin, $W = \{w \in \mathbb{R}^n \mid \Sigma_{i=1}^n c_i w_i = 0\}$. We note that $W^{\perp} = \mathbf{c} = (c_1, \ldots, c_n)^T$, is the normal vector of the (hyper-)plane W. Then the orthogonal projection of a vector $\mathbf{x} \in \mathbb{R}^n$ onto $W^{\perp}$ is given by the orthogonal projection matrix Q, which is

$$\mathbf{x}' = Q\mathbf{x}, \text{ where } Q = \mathbf{c}(\mathbf{c}^T \mathbf{c})^{-1} \mathbf{c}^T = (\mathbf{c}^T\mathbf{c})^{-1} \mathbf{c}\mathbf{c}^T, \text{ since } \mathbf{c}^T\mathbf{c} \in \mathbb{R}.$$

Now, if we scale $\mathbf{c}$ so that $\mathbf{c}$ is a unit vector, then we have

$$Q = \mathbf{c}\mathbf{c}^T.$$

Now, let $\mathbf{x}^*$ be the image of the reflection of $\mathbf{x}$ across W, then the distance between $\mathbf{x}$ and $\mathbf{x}^*$ is twice as much as the distance between $\mathbf{x}$ and $\mathbf{x}'$. Hence, we have

$$\mathbf{x}^* = \mathbf{x} - 2\mathbf{x}' = (\mathbb{I} - 2Q)\mathbf{x} = (\mathbb{I} - 2\mathbf{c}\mathbf{c}^T)\mathbf{x},$$

where Q is the matrix representation of orthogonal projection onto W. The matrix

$H_W = \mathbb{I} - 2\mathbf{c}\mathbf{c}^T$, where $\mathbf{c}$ is a unit normal vector for the plane W is called the reflection matrix for the plane W and sometimes is also called the Householder matrix.

EXAMPLE 4.4.2

Compute the reflection of the vector $\mathbf{x} = (-1, 3, -4)$ across the plane $2x - y + 7z = 0$.

To answer the question, we let $\mathbf{c} = \dfrac{(2,-1,7)^T}{\sqrt{2^2 + (-1)^2 + 7^2}} = \dfrac{(2,-1,7)^T}{\sqrt{54}}$, then

$$H_W = \mathbb{I} - 2\mathbf{c}\mathbf{c}^T = \mathbb{I} - \frac{(2,-1,7)^T(2,-1,7)}{27} = \begin{bmatrix} 23/27 & 2/27 & -14/27 \\ 2/27 & 26/27 & 7/27 \\ -14/27 & 7/27 & -22/27 \end{bmatrix}$$

Hence, the image of reflection is $H_W\mathbf{x} = (39/27, 48/27, 123/27)^T$.

4.5 Properties of Symmetric Matrices

Definition 4.5.1

Let $\langle , \rangle = \cdot$ be the standard inner product on $\mathbb{R}^n$. The real $n \times n$ matrix A is **symmetric** if and only if

$$\langle A\mathbf{x}, \mathbf{y}\rangle = \langle \mathbf{x}, A\mathbf{y}\rangle \ \forall \mathbf{x}, \mathbf{y} \in \mathbb{R}^n.$$

Since this Definition is independent of the choice of basis, symmetry is a property that depends only on the linear operator A and a choice of inner product.

The above Definition suggests that a symmetric matrix is a square matrix that is equal to its transpose. Formally, matrix A is symmetric if

$$A = A^T.$$

Because if $A = A^T$, then

$$\langle A\mathbf{x}, \mathbf{y}\rangle = (A\mathbf{x})^T\mathbf{y} = \mathbf{x}^T A^T \mathbf{y} = \mathbf{x}^T A\mathbf{y} = \langle \mathbf{x}, A\mathbf{y}\rangle.$$

On the other hand, if $\langle A\mathbf{x}, \mathbf{y}\rangle = \langle \mathbf{x}, A\mathbf{y}\rangle$, then let $\mathbf{x} = \mathbf{e}_i$ and $\mathbf{y} = \mathbf{e}_j$, then

$$\langle A\mathbf{x}, \mathbf{y}\rangle = \langle A\mathbf{e}_i, \mathbf{e}_j\rangle = \langle A_i, \mathbf{e}_j\rangle = A_{ji},\ A_i \text{ is the } i\text{-th column of } A,$$

$$\langle \mathbf{x}, A\mathbf{y}\rangle = \langle \mathbf{e}_i, A\mathbf{e}_j\rangle = \langle \mathbf{e}_i, A_j\rangle = A_{ij},\ A_j \text{ is the } j\text{-th column of } A,$$

thus,

$$\langle A\mathbf{x}, \mathbf{y}\rangle = \langle \mathbf{x}, A\mathbf{y}\rangle \Leftrightarrow A_{ij} = A_{ji} \Leftrightarrow A^T = A.$$

Symmetric matrices play an important role in a broad range of applications. Not only are the eigenvalues of a symmetric matrix necessarily real, the eigenvectors always form an orthogonal basis. Hence it is the most common topic to study in terms of orthogonal bases.

To introduce the next theorem, we first define a Hermitian matrix.

Definition 4.5.2

A **Hermitian matrix** or **self-adjoint matrix** is a complex square matrix that is equal to its own conjugate transpose, that is, the element in the i-th row and j-th column is equal to the complex conjugate of the element in the j-th row and i-th column, for all indices i and j:

$$a_{ij} = \overline{a_{ji}} \quad \text{or} \quad A = \overline{A^T}$$

Hermitian matrices can be understood as the complex extension of real symmetric matrices. If the conjugate transpose of a matrix A is denoted by A^*, then the Hermitian property can be written concisely as

$$A = A^*.$$

Lemma 4.5.1 If A is a symmetric matrix, $\mathbf{x}$ is an eigenvector of A, and $\mathbf{y}$ is orthogonal to $\mathbf{x}$, then

$$\mathbf{y} \perp A\mathbf{x}, \quad A\mathbf{x} \perp \mathbf{x}, \quad A\mathbf{y} \perp A\mathbf{x}$$

Proof: To show $\mathbf{y} \perp A\mathbf{x}$, note that $\mathbf{x} \cdot \mathbf{y} = 0$, and

$$A\mathbf{x} \cdot \mathbf{y} = l\mathbf{x} \cdot \mathbf{y} = 0.$$

To show $A\mathbf{y} \perp \mathbf{x}$,

$$A\mathbf{y} \cdot \mathbf{x} = (A\mathbf{y})^T\mathbf{x} = \mathbf{y}^T A^T \mathbf{x} = \mathbf{y}^T(A\mathbf{x}) = \mathbf{y}^T\lambda\mathbf{x} = \lambda\mathbf{y} \cdot \mathbf{x} = 0.$$

Finally,

$$A\mathbf{y} \cdot A\mathbf{x} = (A\mathbf{y})^T(A\mathbf{x}) = \mathbf{y}^T A^T \lambda\mathbf{x} = \lambda\mathbf{y}^T(A\mathbf{x}) = \lambda\mathbf{y}^T\lambda\mathbf{x} = \lambda^2\mathbf{y} \cdot \mathbf{x} = 0.$$

Theorem 4.5.1

Let $A = A^T$ be a real symmetric $n \times n$ matrix. Then

1. All the eigenvalues of A are real.
2. Eigenvectors corresponding to distinct eigenvalues are orthogonal.
3. There is an orthonormal basis of $\mathbb{R}^n$ consisting of n eigenvectors of A.

Proof: First, if $A = A^T$ is real, symmetric, then

$$A\mathbf{v} \cdot \mathbf{w} = \mathbf{v} \cdot A\mathbf{w}.$$

Note in $\mathbb{R}^n$, $\cdot$ is the regular dot product, and in $\mathbb{C}^n$ we have

$$\mathbf{v} \cdot \mathbf{w} = \mathbf{v}^T \overline{\mathbf{w}} \cdot$$

To prove property (1), we note that if $\lambda \in \mathbb{C}$ then

$$A\mathbf{v} \cdot \mathbf{v} = \mathrm{l}\mathbf{v} \cdot \mathbf{v} = \lambda|\mathbf{v}|^2.$$

On the other hand,

$$\mathbf{v} \cdot {}^{A}\mathbf{v} = \mathbf{v} \cdot \lambda\mathbf{v} = \mathbf{v}^T \overline{\lambda\mathbf{v}} = \lambda|\mathbf{v}|^2 .$$

Hence

$$\lambda|\mathbf{v}|^2 = \overline{\lambda}|\mathbf{v}|^2 \Rightarrow \lambda = \overline{\lambda}$$

Thus $\lambda \in \mathbb{R}$.

To prove (2), suppose

$$A\mathbf{v} = a\mathbf{v}, \quad A\mathbf{w} = b\mathbf{w}, \quad a \neq b$$

Then

$$a\mathbf{v} \cdot \mathbf{w} = A\mathbf{v} \cdot \mathbf{w} = \mathbf{v} \cdot A\mathbf{w} = \mathbf{v} \cdot b\mathbf{w} = b\mathbf{v} \cdot \mathbf{w}.$$

This in turn implies $(a - b)(\mathbf{v} \cdot \mathbf{w}) = 0$. Since $a \neq b$, we must have that

$$\mathbf{v} \cdot \mathbf{w} = 0.$$

Thus distinct eigenvalues correspond to orthogonal eigenvectors.

To prove the final statement, if all the eigenvalues of A are distinct, then the corresponding eigenvectors are orthogonal and they form a basis. By normalization, we obtain an orthonormal basis.

To prove the general case, we proceed by induction on the size n of the matrix A. Let $T : \mathbb{R}^n \to \mathbb{R}^n$ be the linear transformation defined by $T\mathbf{x} = A\mathbf{x}$.

To start, the case of a 1×1 matrix is trivial. Suppose A has size $n \times n$. We know that A has at least one real eigenvalue, λ_1, and $\mathbf{v}_1$ is the associated eigenvector. Let $V = \mathrm{Span}(\mathbf{v}_1)$, and

$$V^{\perp} = \{\mathbf{w} \in \mathbb{R}^{\mathrm{n}} | \mathbf{v}_1 \cdot \mathbf{w} = 0\}$$

denote the orthogonal complement to V.

We note that A defines a linear transformation on $V^{\perp}$ since for any $\mathbf{w} \in V^{\perp}$, then by Lemma 4.5.1, $\mathbf{v}_1 \cdot A\mathbf{w} = A\mathbf{v}_1 \cdot \mathbf{w} = \lambda_1\mathbf{v}_1 \cdot \mathbf{w} = 0$. Thus, A restricted to $V^{\perp}$ is a linear transformation from $V^{\perp}$ to $V^{\perp}$. Let $\mathbf{y}_1, \ldots, \mathbf{y}_{n-1}$ be an orthonormal basis for $V^{\perp}$, then $\mathcal{B} = \{\mathbf{v}_1/|\mathbf{v}_1|, \mathbf{y}_1, \ldots, \mathbf{y}_{n-1}\}$ is an orthonormal basis for $\mathbb{R}^n$. The matrix representation of T relative to the basis $\mathcal{B}$ is a block matrix, $\begin{bmatrix} \lambda_1 & 0 \\ 0 & A' \end{bmatrix}$. The first column is $[T(\mathbf{v}_1)]\mathcal{B}$, the coordinate vector of the image of the linear transformation T on $\mathbf{v}_1$. By Lemma 4.5.3, for any $\mathbf{y}_j \in V^{\perp}$, $T(\mathbf{y}_j) = A\mathbf{y}_j \perp \mathbf{v}_1$, hence the first row must be $[\lambda, 0, \ldots, 0]$. Moreover, $A' = (a'_{ij})$ is symmetric, since

$$a'_{ij} = T(\mathbf{y}_j)\cdot \mathbf{y}_i = (A\mathbf{y}_j)\cdot \mathbf{y}_i = (A\mathbf{y}_j)^T \mathbf{y}_i = \mathbf{y}_i^T (A^T \mathbf{y}_i)$$
$$= \mathbf{y}_i^T(\mathbf{y}_i) = \mathbf{y}_j \cdot T(\mathbf{y}_j) = T(\mathbf{y}_j)\cdot \mathbf{y}_i = a'_{ji}$$

Thus, A' is an $(n-1)\times(n-1)$ symmetric matrix, and by induction hypothesis, the eigenvectors of A' form an orthonormal basis $\mathbf{u}_2, \ldots, \mathbf{u}_n$ for $V^{\perp}$. Then appending the unit eigenvector $\mathbf{u}_1 = \mathbf{v}_1/|\mathbf{v}_1|$, we obtain the orthonormal basis for $\mathbb{R}^n$ consisting of n eigenvectors of A.

The orthonormal eigenvector basis serves to diagonalize the symmetric matrix, resulting in the following spectral factorization formula.

Theorem 4.5.2

(Spectral decomposition) Let A be a real, symmetric matrix. Then there exists an orthogonal matrix Q such that

$$A = Q\Lambda Q^{-1} = Q\Lambda Q^T,$$

where $\Lambda = \text{diag}(\lambda_1, \ldots, \lambda_n)$ is a real diagonal matrix. The eigenvalues of A appear on the diagonal of Λ, while the columns of Q are the corresponding orthonormal eigenvectors.

Proof: From the proof of the previous theorem, we can construct a basis $\mathcal{B}$ for $\mathbb{R}^n$ consisting of ^{n} orthonormal eigenvectors of A. Let Q be a matrix whose columns are the basis elements of $\mathcal{B}$. Let $T : \mathbb{R}^n \to \mathbb{R}^n$ be the linear transformation defined by $T\mathbf{x} = A\mathbf{x}$. Then the matrix representation of T relative to $\mathcal{B}$ is a diagonal matrix Λ with eigenvalues being the diagonal entries. Q is a change of basis matrix such that $A = [T]_{\varepsilon} = Q\Lambda Q^{-1}$ where ε denotes the standard basis.

Definition 4.5.3

An $n \times n$ symmetric matrix A is called positive definite if it satisfies $\mathbf{x}^T A\mathbf{x} > 0\ \forall \mathbf{0} \neq \mathbf{x} \in \mathbb{R}^n$.

More generally, an $n \times n$ Hermitian matrix A is said to be positive definite if $0 < \mathbf{x}^* A\mathbf{x} \in \mathbb{R}$, $\forall \mathbf{0} \neq \mathbf{x} \in \mathbb{C}^n$, $\mathbf{x}^*$ denotes the conjugate transpose of $\mathbf{x}$.

The negative definite, positive semi-definite, and negative semi-definite matrices are defined in the same way, i.e., the expression $\mathbf{x}^*\mathbf{A}\mathbf{x}$ is required to be always negative, non-negative, and non-positive, respectively.

Theorem 4.5.3

A symmetric matrix $A = A^T$ is positive definite if and only if all of its eigenvalues are strictly positive.

Proof: ($\Rightarrow$) Let λ be an eigenvalue with corresponding eigenvector $\mathbf{v}$, then

$$0 < \mathbf{x}^T A\mathbf{x} = \mathbf{v} \cdot A\mathbf{v} = \mathbf{v} \cdot \lambda\mathbf{v} = \lambda|\mathbf{v}|^2.$$

Hence we must have that $\lambda > 0$.

($\Leftarrow$) On the other hand, if $\lambda_i > 0$ for all the eigenvalues, then let $\mathbf{v}_1, \ldots, \mathbf{v}_n$ be a orthonormal basis in $\mathbb{R}^n$ consisting of the eigenvectors of A, then

$$X^T Ax = \left(\sum_{i=1}^{n} c_i\mathbf{v}_i\right)^T A\left(\sum_{i=1}^{n} c_i\mathbf{v}_i\right) = \left(\sum_{i=1}^{n} c_i\mathbf{v}_i\right)^T \left(\sum_{i=1}^{n} c_i A\mathbf{v}_i\right)$$

$$= \left(\sum_{i=1}^{n} c_i\mathbf{v}_i\right)^T \left(\sum_{i=1}^{n} c_i\lambda_i\mathbf{v}_i\right) = \sum_{i=1}^{n} \lambda_i c_i^2 > 0, \quad \text{since } \mathbf{x} \neq \mathbf{0}, c_i \neq 0$$

4.6 QR Factorization

In linear algebra, a **QR decomposition**, or **QR factorization** of a square matrix is a decomposition of a matrix A into a product $A = QR$ of an orthogonal matrix Q and an upper triangular matrix R. If A is a real matrix, then Q is an orthogonal matrix, i.e., $QTQ = \mathbb{I}$, and if A is a complex matrix, then Q is a **unitary matrix**, i.e., $Q^*Q = \mathbb{I}$).

In general, when A is an $m \times n$ matrix, with $m \geq n$, as the product of an $m \times m$ unitary matrix Q and an m x n upper triangular matrix R,

$$A = QR = Q\begin{bmatrix} R_1 \\ \mathbb{O} \end{bmatrix}$$

where R_1 is an $n \times n$ upper triangular matrix, $\mathbb{O}$ is an $(m - n) \times n$ zero matrix.

The QR factorization can be obtained by the Gram-Schmidt process applied to the columns of the full column rank matrix $A = [\mathbf{a}_1, \ldots, \mathbf{a}_n]$.

Let inner product $\langle\mathbf{v}, \mathbf{w}\rangle = \mathbf{v}^{\mathsf{T}}\mathbf{w}$ in the real case, or $\langle\mathbf{v}, \mathbf{w}\rangle = \mathbf{v}^*\mathbf{w}$ for the complex case.

Recall the projection:

$$\text{proj}_{\mathbf{e}}\mathbf{a} = \frac{\langle \mathbf{e}, \mathbf{a} \rangle}{\langle \mathbf{e}, e \rangle}\mathbf{e}$$

then:

$$\begin{aligned}
&\mathbf{u}_1 = \mathbf{a}_1, & \mathbf{e}_1 &= \frac{\mathbf{u}_1}{|\mathbf{u}_1|} \\
&\mathbf{u}_2 = \mathbf{a}_2 - \text{proj}_{\mathbf{u}_1}\mathbf{a}_2, & \mathbf{e}_2 &= \frac{\mathbf{u}_2}{|\mathbf{u}_2|} \\
&\mathbf{u}_3 = \mathbf{a}_3 - \text{proj}_{\mathbf{u}_1}\mathbf{a}_3 - \text{proj}_{\mathbf{u}_2}\mathbf{a}_3, & \mathbf{e}_3 &= \frac{\mathbf{u}_3}{|\mathbf{u}_3|} \\
&\vdots & &\vdots \\
&\mathbf{u}_n = \mathbf{a}_n - \sum_{j=1}^{n-1}\text{proj}_{\mathbf{u}_j}\mathbf{a}_n, & \mathbf{e}_n &= \frac{\mathbf{u}_n}{|\mathbf{u}_n|}
\end{aligned}$$

Express the columns $\mathbf{a}_i$, $i = 1, \dots, n$, over our newly computed orthonormal basis:

$$\begin{aligned}
\mathbf{a}_1 &= \langle \mathbf{e}_1, \mathbf{a}_1 \rangle \mathbf{e}1 \\
\mathbf{a}_2 &= \langle \mathbf{e}_1, \mathbf{a}_2 \rangle \mathbf{e}_1 + \langle \mathbf{e}_2, \mathbf{a}_2 \rangle \mathbf{e}_2 \\
\mathbf{a}_3 &= \langle \mathbf{e}_1, \mathbf{a}_3 \rangle \mathbf{e}_1 + \langle \mathbf{e}_2, \mathbf{a}_3 \rangle \mathbf{e}_2 + \langle \mathbf{e}_3, \mathbf{a}_3 \rangle \mathbf{e}_3 \\
&\vdots \\
\mathbf{a}_n &= \sum_{j=1}^{n} \langle \mathbf{e}_j, \mathbf{a}_n \rangle \mathbf{e}_j
\end{aligned}$$

where

$$\langle \mathbf{e}_i, \mathbf{a}_i \rangle = |\mathbf{u}_i|.$$

Write this in a matrix form $A = QR$ where:

$$Q = [\mathbf{e}_1, \dots, \mathbf{e}_n] \quad \text{and} \quad R = \begin{pmatrix} \langle \mathbf{e}_1, \mathbf{a}_1 \rangle & \langle \mathbf{e}_1, \mathbf{a}_2 \rangle & \langle \mathbf{e}_1, \mathbf{a}_3 \rangle & \cdots \\ 0 & \langle \mathbf{e}_1, \mathbf{a}_2 \rangle & \langle \mathbf{e}_1, \mathbf{a}_3 \rangle & \cdots \\ 0 & 0 & \langle \mathbf{e}_3, \mathbf{a}_3 \rangle & \cdots \\ \vdots & \vdots & \vdots & \ddots \end{pmatrix}$$

EXAMPLE 4.6.1

Consider the matrix $A = \begin{bmatrix} 1 & 1 & 0 \\ 1 & 0 & 1 \\ 0 & 1 & 1 \end{bmatrix}$. Let

$$\mathbf{a}_1 = (1, 1, 0)^T, \mathbf{a}_2 = (1, 0, 1)^T, \mathbf{a}_3 = (0, 1, 1)^T.$$

Performing the Gram-Schmidt procedure, we obtain

$$\mathbf{u}_1 = \mathbf{a}_1 = (1,1,0)^T$$

$$\mathbf{e}_1 = \frac{\mathbf{u}_1}{|\mathbf{u}_1|} = \frac{1}{\sqrt{2}}(1,1,0)^T;$$

$$\mathbf{u}_2 = \mathbf{a}_2 - (\mathbf{a}_2 \cdot \mathbf{e}_1)\mathbf{e}_1 = (1,0,1)^T - \frac{1}{\sqrt{2}} \cdot \frac{1}{\sqrt{2}}(1,1,0)^T = \frac{1}{2}(1,-1,2)^T$$

$$\mathbf{e}_2 = \frac{\mathbf{u}_2}{|\mathbf{u}_2|} = \frac{1}{\sqrt{6}}(1,-1,2)^T;$$

$$\mathbf{u}_3 = \mathbf{a}_3 - (\mathbf{a}_3 \cdot \mathbf{e}_1)\mathbf{e}_1 - (\mathbf{a}_3 \cdot \mathbf{e}_2)\mathbf{e}_2$$

$$= (0,1,1)^T - \frac{1}{2}(1,1,0)^T - \frac{1}{6}(1,-1,2)^T = \frac{2}{3}(-1,1,1)^T$$

$$\mathbf{e}_3 = \frac{\mathbf{u}_3}{|\mathbf{u}_3|} = \frac{1}{\sqrt{3}}(-1,1,1)^T$$

Hence, $A = QR$ where

$$Q = [\mathbf{e}_1, \mathbf{e}_2, \mathbf{e}_3] = \begin{bmatrix} \frac{1}{\sqrt{2}} & \frac{1}{\sqrt{6}} & -\frac{1}{\sqrt{3}} \\ \frac{1}{\sqrt{2}} & -\frac{1}{\sqrt{6}} & \frac{1}{\sqrt{3}} \\ 0 & \frac{2}{\sqrt{6}} & \frac{1}{\sqrt{3}} \end{bmatrix}$$

$$R = \begin{pmatrix} \langle \mathbf{e}_1, \mathbf{a}_1 \rangle & \langle \mathbf{e}_1, \mathbf{a}_2 \rangle & \langle \mathbf{e}_1, \mathbf{a}_3 \rangle \\ 0 & \langle \mathbf{e}_2, \mathbf{a}_2 \rangle & \langle \mathbf{e}_2, \mathbf{a}_3 \rangle \\ 0 & 0 & \langle \mathbf{e}_3, \mathbf{a}_3 \rangle \end{pmatrix} = \begin{bmatrix} \sqrt{2} & \frac{1}{\sqrt{2}} & \frac{1}{\sqrt{2}} \\ 0 & \frac{3}{\sqrt{6}} & \frac{1}{\sqrt{6}} \\ 0 & 0 & \frac{2}{\sqrt{3}} \end{bmatrix}$$

Two alternative technique can be applied for QR decomposition.

Givens rotation: For 2-dimensional vector $\mathbf{v} = (v_1, v_2)^T$, one can find a rotation matrix $G_2 = \begin{bmatrix} \cos(\theta) & -\sin(\theta) \\ \sin(\theta) & \cos(\theta) \end{bmatrix}$ such that $G_2(v_1, v_2)^T = \left(\sqrt{v_1^2 + v_2^2}, 0\right)^T$

where $\cos(\theta) = \dfrac{v_1}{\sqrt{v_1^2 + v_2^2}}$ and $\sin(\theta) = \dfrac{v_2}{\sqrt{v_1^2 + v_2^2}}$. For a m-dimensional vector $\mathbf{v} = (v_1, \ldots, v_i, \ldots, v_m)^T$, a rotation matrix

$$G_m(i,j) = \begin{bmatrix} 1 & \cdots & 0 & \cdots & 0 & \cdots & 0 \\ \vdots & \ddots & \vdots & & \vdots & & \vdots \\ 0 & \cdots & c & \cdots & -s & \cdots & 0 \\ \vdots & & \vdots & \ddots & \vdots & & \vdots \\ 0 & \cdots & s & \cdots & c & \cdots & 0 \\ \vdots & & \vdots & & \vdots & \ddots & \vdots \\ 0 & \cdots & 0 & \cdots & 0 & \cdots & 1 \end{bmatrix}$$

Will rotate v to a vector whose j-th element is zero, i.e.,

$$G_m(i,j)\mathbf{v} = \left(v_1, \ldots, \sqrt{v_i^2 + v_2^2}, \ldots, 0, \ldots, v_m\right)^T$$

Notice that $G_m(i,j)$ is an orthogonal matrix. Rotate each column of the $m \times n$ matrix A such that

$$\prod_{j=n+1}^{m} G_m(n,j) \cdots \prod_{j=3}^{m} G_m(2,j) \prod_{j=2}^{m} G_m(1,j) A = \begin{bmatrix} R_1 \\ \mathbb{O} \end{bmatrix}$$

where R_1 is an $n \times n$ upper triangular matrix, $\mathbb{O}$ is an $(mn) \times n$ zero matrix.

Let

$$Q = \left(\prod_{j=n+1}^{m} G_m(n,j) \cdots \prod_{j=3}^{m} G_m(2,j) \prod_{j=2}^{m} G_m(1,j) \right)^{-1}$$

and $R = \begin{bmatrix} R_1 \\ \mathbb{O} \end{bmatrix}$, then QR is a QR decomposition of A.

Householder transformation: Householder transformation is a linear transformation that describes a reflection about a plane or hyperplane containing the origin. For any two vectors $\mathbf{v}$, $\mathbf{w}$ with same length, $\mathbf{v}$, $\mathbf{w}$ are reflections of each other with respect to a hyperplane which passes through the origin and is orthogonal to the unit vector $\mathbf{u} = \dfrac{\mathbf{v} - \mathbf{w}}{|\mathbf{v} - \mathbf{w}|}$. In fact, the matrix $P = \mathbb{I} - 2\mathbf{u}\mathbf{u}^*$ will define a linear transformation such that $\mathbf{w} = P\mathbf{v}$ and P is called a Householder matrix having the following properties:

1. $P^* = P$;
2. $P^*P = \mathbb{I}$;
3. $P2 = \mathbb{I}$;
4. diag($\mathbb{I}$, P) is also a household matrix;
5. det(P) = −1.

To find a QR decomposition of A, we construct a Householder matrix P_1 where $\mathbf{v}$ is the first column of A, and $\mathbf{w} = {}_P\mathbf{v}$ is a column $[\tilde{a}_{11}, 0, \ldots, 0]^T$ such that

$$P_1 A = \begin{bmatrix} \tilde{a}_{11} & A^{(1)}_{1\times n-1} \\ \mathbf{0}_{n-2\times 1} & \overline{A}^{(1)}_{n-1\times n-1} \end{bmatrix}$$

Construct a Householder matrix $\tilde{P}_2$ for the first column of $\overline{A}^{(1)}_{n-1\times n-1}$ such that

$$\overline{P}_2 \overline{A}^{(1)}_{n-1\times n-1} = \begin{bmatrix} \tilde{a}_{22} & \overline{A}^{(1)}_{1\times n-2} \\ \mathbf{0}_{n-2\times 1} & \overline{A}^{(2)}_{n-2\times n-2} \end{bmatrix}$$

Let $P_2 = \begin{bmatrix} 1 & \mathbf{0}_{1\times n-1} \\ \mathbf{0}_{n-1\times 1} & \tilde{P}_2 \end{bmatrix}$, we have

$$P_2 P_1 A = \begin{bmatrix} \tilde{a}_{11} & \tilde{a}_{12} & \overline{A}^{(1)}_{1\times n-2} \\ 0 & \tilde{a}_{22} & \overline{A}^{(2)}_{1\times n-2} \\ \mathbf{0}_{n-2\times 1} & \mathbf{0}_{n-2\times 1} & \overline{A}^{(2)}_{n-2\times n-2} \end{bmatrix}$$

Repeat the constructions for Householder matrices and get $P_n \ldots P_1 A = \begin{bmatrix} R_1 \\ \mathbb{O} \end{bmatrix} = R$. Then $Q = (P_n \ldots P_1)^{-1} = P_1 \ldots P_n$ and R form a QR decomposition of A.

4.7 Singular Value Decomposition

In linear algebra, the singular value decomposition (SVD) is a factorization of a real or complex matrix. It is a generalization of the spectral factorization. It has many useful applications.

Definition 4.7.1

A non-negative real number a is a **singular value** for an $m \times n$ matrix M if and only if there exist unit-length vectors **u** in $\mathbb{F}^m$ and $\mathbf{v} \in \mathbb{F}^n$ such that

$$M\mathbf{v} = \sigma\mathbf{u} \quad \text{and} \quad M^*\mathbf{u} = \sigma\mathbf{v}$$

The vectors **u** and **v** are called left-singular and right-singular vectors for σ, respectively.

Definition 4.7.2

The singular value decomposition (SVD) of an $m \times n$ real or complex matrix M is a factorization of the form

$$M = U\Sigma V^*,$$

where U is an $m \times m$ real or complex unitary matrix, $\Sigma = \text{diag}(\sigma_1, \ldots, \sigma_k)$ where $\sigma_1 \geq \ldots \geq \sigma_k \geq 0$, $k = \min(m, n)$, is an $m \times n$ rectangular diagonal matrix with non-negative real numbers on the diagonal, and V is an $n \times n$ real or complex unitary matrix. The diagonal entries σ_i of Σ are known as the singular values of M The columns of U and the columns of V are called the left-singular vectors and right-singular vectors of M.

Remark 4.7.1 Let M be a real $m \times n$ matrix. Let $U = [\mathbf{u}_1, \ldots, \mathbf{u}_m]$ and $V = [\mathbf{v}_1, \ldots, \mathbf{v}_n]$. From the relations

$$M\mathbf{v}_j = \sigma_j \mathbf{u}_j; \quad M^T \mathbf{u}_j = \sigma_j \mathbf{v}_j; \quad j = 1, \ldots, k$$

it follows that

$$M^T M\mathbf{v}_i = \sigma_i^2 \mathbf{v}_i.$$

Hence, the squares of the singular values are the eigenvalues of M^TM, which is a symmetric matrix.

Remark 4.7.2 If $M \in \mathbb{R}^{m\times n}$, then M^TM is symmetric positive semi-definite matrix, and the singular values of M are defined to be the square roots of the eigenvalues of M^TM. The singular values of M will be denoted by $\sigma_1, \ldots, \sigma_n$. It is customary to list the singular values in decreasing order so it will be assumed that $\sigma_1 \geq \ldots \geq \sigma_n$.

The singular value decomposition (SVD) can also be written as

$$M = \sigma_1\mathbf{u}_1\mathbf{v}_1^T + \cdots + \sigma_m\mathbf{u}_m\mathbf{v}_m^T$$

You should see a similarity between the singular value decomposition and the spectral decomposition stated in Theorem 4.5.2. In fact, if M is symmetric, then SVD and spectral decomposition are the same. In general, singular value decomposition of a matrix is not unique. The right singular vectors are orthonormal eigenvectors of M^TM. If an eigenspace of this matrix is 1-dimensional there are two choices for the corresponding singular vector, these choices are negatives of each other. If an eigenspace has dimension greater than one then there are infinitely many choices for the orthonormal eigenvectors, but any of these choices would be an orthonormal basis of the same eigenspace.

EXAMPLE 4.7.5

Find the SVD of M, where $M = \begin{bmatrix} 3 & 2 & 2 \\ 2 & 3 & -2 \end{bmatrix}$.

Proof: First, we compute the singular values by finding the eigenvalues of MM^T.

$$MM^T = \begin{bmatrix} 17 & 8 \\ 8 & 17 \end{bmatrix}, \ \det(MM^T - \lambda \mathbb{I}) = 12 - 34\lambda + 225 = (\lambda - 25)(\lambda - 9)$$

$$\sigma_1 = \sqrt{25} = 5, \quad \sigma_2 = \sqrt{9} = 3$$

Second, we will find the right singular vectors by finding an orthonormal set of eigenvectors of M^TM.

$$M^TM = \begin{bmatrix} 13 & 12 & 2 \\ 12 & 13 & -2 \\ 2 & -2 & 8 \end{bmatrix}.$$

The eigenvalues of M^TM are 25, 9, and 0. Moreover, M^TM is symmetric, so the eigenvectors are orthogonal. Now to compute the eigenvectors of M^TM corresponding to the eigenvalues 25, 9, 0 are

$$\mathbf{v}_1 = \begin{bmatrix} 1/\sqrt{2} \\ 1/\sqrt{2} \\ 0 \end{bmatrix}, \quad \mathbf{v}_2 = \begin{bmatrix} 1/\sqrt{18} \\ -1/\sqrt{18} \\ 4/\sqrt{18} \end{bmatrix}, \quad \mathbf{v}_3 = \begin{bmatrix} 2/3 \\ -2/3 \\ -1/3 \end{bmatrix},$$

Hence so far we have

$$M = U\Sigma V^T = U\begin{bmatrix} 5 & 0 & 0 \\ 0 & 3 & 0 \end{bmatrix}\begin{bmatrix} 1/\sqrt{2} & 1/\sqrt{2} & 0 \\ 1/\sqrt{18} & -1/\sqrt{18} & 4/\sqrt{18} \\ 2/3 & -2/3 & -1/3 \end{bmatrix}$$

To find U, we realize that $M\mathbf{v}_i = \sigma_i \mathbf{u}_i$, thus $\mathbf{u}_i = \frac{1}{\sigma_i} M\mathbf{v}_i$.

$$U = [\mathbf{u}_1 \ \mathbf{u}_2] = \begin{bmatrix} 1/\sqrt{2} & 1/\sqrt{2} \\ 1/\sqrt{2} & -1/\sqrt{2} \end{bmatrix}$$

Thus, we have the final SVD of M:

$$M = U\Sigma V^T = \begin{bmatrix} 1/\sqrt{2} & 1/\sqrt{2} \\ 1/\sqrt{2} & -1/\sqrt{2} \end{bmatrix} \begin{bmatrix} 5 & 0 & 0 \\ 0 & 3 & 0 \end{bmatrix} \begin{bmatrix} 1/\sqrt{2} & 1/\sqrt{2} & 0 \\ 1/\sqrt{12} & -1/\sqrt{18} & 4/\sqrt{18} \\ 2/3 & -2/3 & -1/3 \end{bmatrix}$$

Mathematica [Wol15], Maple [Map16], MATLAB [TM], Sage [SJ05], and many other computer software programs can compute matrix decompositions. Maple [Map16] is a symbolic and numeric computing environment, and is also a multi-paradigm programming language developed by Maplesoft. Below, we will provide an example to illustrate how to use Maple to perform SVD. We first create a matrix A, and then perform singular value decomposition via Maple.

```
>A:=Matrix(3,4,(i,j)->i+j-1);
```

$$A := \begin{bmatrix} 1 & 2 & 3 & 4 \\ 2 & 3 & 4 & 5 \\ 3 & 4 & 5 & 6 \end{bmatrix}$$

```
>SingularValues(A,output=['S','U','Vt']);
```

$$\begin{bmatrix} 13.0111937212366 \\ 0.841925144210536 \\ 5.034905542\times10^{-17} \end{bmatrix}, \begin{bmatrix} -0.417672938745049 & -0.811715867513632 & 0.408248290463862 \\ -0.564727113803818 & -0.120063231146631 & -0.816496580927726 \\ -0.711781288862587 & 0.57155774055220368 & 0.408248290463863 \end{bmatrix}$$

$$\begin{bmatrix} -0.283023303767278 & -0.413232827790139 & -0.543442351813000 & -0.673651875835862 \\ 0.787335893710336 & 0.359497746914449 & -0.0683403998814394 & -0.496178546677328 \\ 0.409743990033253 & -0.817239269719659 & 0.405246569339561 & 0.00224871034684536 \\ 0.363469204516187 & -0.179220467659466 & -0.731966678229628 & 0.547717941372907 \end{bmatrix}$$

4.8 Applications of SVD and QR Decompositions

We give two applications of SVD and QR decompositions.

4.8.1 Image Processing

In image processing, we can represent digital images as matrices. Consider a gray scale image having $m \times n$ pixels:, it can be represented as an $m \times n$ matrix. For color images, we need three numbers per pixel, for each color: red, green, and blue (RGB). Then the image processing turns to matrix analysis. For instance, we can easily rotate an image and add two images. Here we give a classic image compression method based on singular value decomposition (SVD).

Let M be an $m \times n$ matrix of an image; there is a factorization of the form

$$M = U\Sigma V^*,$$

where U is an $m \times m$ real or complex unitary matrix, $\Sigma = \text{diag}(\sigma_1, \ldots, \sigma_k)$ where $\sigma_1 \geq \ldots \geq \sigma_k \geq 0$, $k = \min(m, n)$, is an $m \times n$ rectangular diagonal matrix with non-negative real numbers on the diagonal, and V is an $n \times n$ real or complex unitary matrix. Let $U = [\mathbf{u}_1, \ldots, \mathbf{u}_m]$ and $V = [\mathbf{v}_1, \ldots, \mathbf{v}_n]$. The singular value decomposition (SVD) of M can also be written as

$$M = \sigma_1 \mathbf{u}_1 \mathbf{v}_1^T + \cdots + \sigma_m \mathbf{u}_m \mathbf{v}_m^T$$

Following the fact $\sigma_1 \geq \ldots \geq \sigma_k \geq 0$, most of the information of the image is included in terms which have greater singular values. This motives people to compress the image by dropping the terms with smaller singular values. For instance, we only need to store $l(m + n + 1)$ numbers if we use

$$\tilde{M} = \sigma_1 \mathbf{u}_1 \mathbf{v}_1^T + \cdots + \sigma_i \mathbf{u}_i \mathbf{v}_i^T$$

instead of storing mn numbers by using matrix M. We compare the results for different l with an example. The image with 512×512-pixels shown at the left in Fig. 4.1 is a photo of one of the authors and the right image in Fig. 4.1 is the compressed one with $l = 1$. The left image in Fig. 4.2 is the compressed one with $l = 50$ and the right one is with $l = 100$.

Note, in terms of storage, the original image requires $mn = 512 \cdot 512 = 262144$ numbers. If $l = 100$, then the storage for compressed image is $l(m + n + 1) = 100(512 + 512 + 1) = 102500$, which is about 39% of the original storage. This saves 60% of storage space, without compromising the quality of the image.

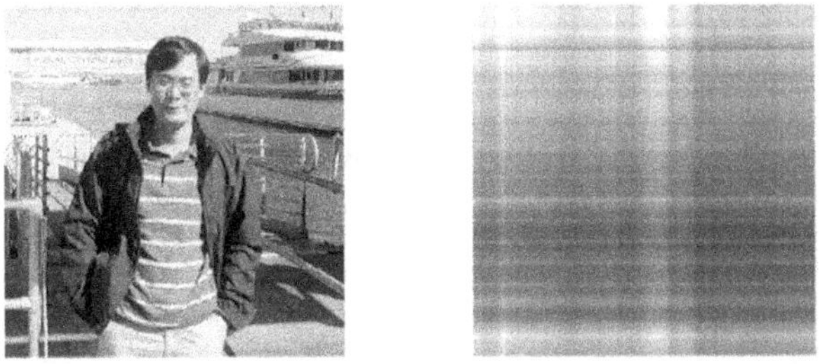

Original image Compressed image with l = 1

Figure 4.1 Image compression

Compressed image with l = 50 Compressed image with l = 100

Figure 4.2 Image compression (cont.)

4.8.2 QR Decomposition and GCD

The matrix can used in some algebraic computations, as a benefit of this situation, we can generalize the computation to numerical way. Let two polynomials $f(x)$ and $g(x)$ as

$$f(x) = f_m x^m + f_{m-1}x^{m-1} + \cdots + f_1 x + f_0,$$
$$g(x) = g_n x^n + g_{n-1}x^{n-1} + \cdots + g_1 x + g_0.$$

The Sylvester matrix of f and g is an $(m+n) \times (m+n)$ matrix consist of the coefficients of f and g.

$$S(f,g) = \begin{bmatrix} f_m & f_{m-1} & \cdots & f_1 & f_0 & 0 & 0 & 0 \\ 0 & f_m & f_{m-1} & \cdots & f_1 & f_0 & 0 & 0 \\ 0 & 0 & \ddots & \ddots & \cdots & \ddots & \ddots & 0 \\ 0 & 0 & 0 & f_m & f_{m-1} & \cdots & f_1 & f_0 \\ g_n & g_{n-1} & \cdots & g_1 & g_0 & 0 & 0 & 0 \\ 0 & g_n & g_{n-1} & \cdots & g_1 & g_0 & 0 & 0 \\ 0 & 0 & \ddots & \ddots & \cdots & \ddots & \ddots & 0 \\ 0 & 0 & 0 & g_n & g_{n-1} & \cdots & g_1 & g_0 \end{bmatrix} \begin{matrix} \left.\vphantom{\begin{matrix}a\\a\\a\\a\end{matrix}}\right\} n \\ \left.\vphantom{\begin{matrix}a\\a\\a\\a\end{matrix}}\right\} m \end{matrix}$$

The Sylvester matrix has the following property [Lai69]:

Theorem 4.8.1

(Suppose the Sylvester matrix has the QR decomposition as $S(f, g) = QR$ where Q is orthogonal and R is upper triangular. Then the last non-zero row of R gives the coefficients of the GCD of f and g. Precisely, let R_d be the last non-zero line of R, then the GCD of f and g can be written as

$$R_d \begin{bmatrix} x^{m+n-1} \\ \vdots \\ x^{m-1} \\ \vdots \\ 1 \end{bmatrix}$$

We can check the property with a simple example. Suppose that

$$f(x) = x^3 + 2x^2 + 2x + 1,\ g(x) = x^2 + 3x + 2,$$

then the Sylvester matrix is

$$S(f,g) = \begin{bmatrix} 1 & 2 & 2 & 1 & 0 \\ 0 & 1 & 2 & 2 & 1 \\ 1 & 3 & 2 & 0 & 0 \\ 0 & 1 & 3 & 2 & 0 \\ 0 & 0 & 1 & 3 & 2 \end{bmatrix}$$

Compute the QR decomposition matrices as

$$Q = \begin{bmatrix} 1/2\sqrt{2} & -1/10\sqrt{10} & 1/2 & -1/20\sqrt{10} & 1/4\sqrt{2} \\ 0 & 1/5\sqrt{10} & 0 & 1/10\sqrt{10} & 1/2\sqrt{2} \\ 1/2\sqrt{2} & 1/10\sqrt{10} & -1/2 & 1/20\sqrt{10} & -1/4\sqrt{2} \\ 0 & 1/5\sqrt{10} & 1/2 & -\dfrac{3\sqrt{10}}{20} & -1/4\sqrt{2} \\ 0 & 0 & 1/2 & 1/4\sqrt{10} & -1/4\sqrt{2} \end{bmatrix}$$

$$R = \begin{bmatrix} \sqrt{2} & 5/2\sqrt{2} & 2\sqrt{2} & 1/2\sqrt{2} & 0 \\ 0 & 1/2\sqrt{10} & \sqrt{10} & \dfrac{7\sqrt{10}}{10} & 1/5\sqrt{10} \\ 0 & 0 & 2 & 3 & 1 \\ 0 & 0 & 0 & -3/5\sqrt{10} & 3/5\sqrt{10} \\ 0 & 0 & 0 & 0 & 0 \end{bmatrix}$$

The last line non-zero line of R is $\begin{bmatrix} 0 & 0 & 0 & 3/5\sqrt{10} & 3/5\sqrt{10} \end{bmatrix}$, by the above theorem, the GCD of $f(x)$ and $g(x)$ should be

$$[0 \quad 0 \quad 0 \quad 3/5\sqrt{10} \quad 3/5 \quad \sqrt{10}] \begin{bmatrix} x^{m+n-1} \\ \vdots \\ x^{m-1} \\ \vdots \\ 1 \end{bmatrix} = 3/5\sqrt{10}(x+1)$$

One can find $GCD(f, g) = x + 1$ which is exactly equal to $3/5\sqrt{10}(x+1)$ up to a non-zero constant scalar.

When small perturbations are applied to the coefficients of the polynomials, the GCD obtained after the perturbation may be different from the original GCD. But for the engineering purposes, the new GCD should be close to the original GCD since only small perturbations are introduced. It is difficult to use the Euclidean algorithm numerically. The QR decomposition focuses on matrix computations and gives a way to compute a numerical approximation. Add perturbations to the two polynomials f and g, we have

$$f(x) = x^3 + 2x^2 + 2x + 1.001,\ g(x) = x^2 + 3x + 1.999.$$

Using the numerical GCD method based on QR factoring [CWZ04], one can find an approximate GCD as

$$0.7069376841584446 + 0.707275837786569x$$
$$= 0.707275837786569(x + 0.999521892859819).$$

Thus, the approximate GCD is $x + 0.999521892859819$.

4.9 Exercises

1. Consider the vectors $\mathbf{a} = (1, 5)$ and $\mathbf{b} = (3, 4)$ in $\mathbb{R}^2$. Find the inner product of these two vectors and find the angle between them.
2. Find k so that $(1, 2, k, 3)$ and $(3, k, 7, 5)$ are orthogonal vectors in $\mathbb{R}^4$.
3. Let W be the subspace of $\mathbb{R}^4$ spanned by $(1, 2, 3, -1)$ and $(2, 4, 7, 2)$. Find a basis of the orthogonal complement $W^\perp$ of W.

4. Let S consist of the following vectors in $\mathbb{R}^4$:
$(1, 1, 0, -1), (1, 2, 1, 3), (1, 1, -9, 2), (16, -13, 1, 3)$.
Is S orthogonal and a basis of $\mathbb{R}^4$? If so, find the coordinate of vector (a, b, c, d) in $\mathbb{R}^4$ relative to the basis ^{S}.

5. Find an orthogonal basis, and an orthonormal basis for the subspace spanned by the following vectors:
$(1, 1, 1, 1), (1, 1, 2, 4), (1, 2, -4, -3)$.

6. Project the vector $\begin{pmatrix} 2 \\ -1 \\ 4 \end{pmatrix}$ orthogonally onto the line $\left\{ c \begin{pmatrix} -3 \\ 1 \\ -3 \end{pmatrix} \middle| c \in \mathbb{R} \right\}$.

7. In $\mathbb{R}^4$, project $\mathbf{v} = \begin{pmatrix} 1 \\ 2 \\ 1 \\ 3 \end{pmatrix}$ onto the line $\left\{ c \cdot \begin{pmatrix} -1 \\ 1 \\ -1 \\ 1 \end{pmatrix} \middle| c \in \mathbb{R} \right\}$

8. Compute the orthogonal projection matrix Q for the subspace $W \subset \mathbb{R}^4$ spanned by $(1, 2, 0, 0)$ and $(1, 0, 1, 1)$. Then find the image of the orthogonal projection of the vector $(3, 2, 1, 5)$ onto W.

9. Compute the reflection matrix for reflection across the plane $x + y - z = 0$. Then find the image of the reflection of the vector $(1, 2, 3)$.

10. Compute the QR decomposition of matrix $A = \begin{bmatrix} -1 & 1 & 4 \\ 1 & 4 & -2 \\ 1 & 4 & 2 \\ 1 & -1 & 0 \end{bmatrix}$

11. Compute the QR decomposition of matrix $A = \begin{bmatrix} 1 & 2 & 0 \\ 0 & 1 & 1 \\ 1 & 0 & 1 \end{bmatrix}$

12. Compute the SVD decomposition of matrix $A = \begin{bmatrix} 3 & 1 \\ 1 & 3 \\ 1 & 1 \end{bmatrix}$

13. Compute the SVD decomposition of matrix $A = \begin{bmatrix} 3 & 1 \\ 1 & 3 \\ 1 & 1 \end{bmatrix}$

14. Compute the SVD decomposition of matrix $A = \begin{bmatrix} 2 & 0 & 8 & 6 & 0 \\ 1 & 6 & 0 & 1 & 7 \\ 5 & 0 & 7 & 4 & 0 \\ 7 & 0 & 8 & 5 & 0 \\ 0 & 1 & 0 & 0 & 7 \end{bmatrix}$

CHAPTER 5

Matrix Decomposition

5.1 Decomposition over $\mathbb{R}$ or $\mathbb{C}$

In linear algebra, a matrix decomposition or matrix factorization is a factorization of a matrix into a product of matrices. There are many different matrix decompositions; each finds use among a particular class of problems. Below, we provide a list of common decompositions based on their applications.

5.1.1 Decompositions to Solve Linear Systems

In general, the following decompositions are commonly used to solve systems of linear equations.

1. LU decomposition: applicable to square matrix A

$$A = LU,$$

where L is lower triangular and U is upper triangular.

2. Rank factorization: applicable to $m \times n$ matrix A of rank r

$$A = CF,$$

where C is an $m \times r$ full column rank matrix and F is an $r \times n$ full row rank matrix.

3. Cholesky decomposition: applicable to square, symmetric, positive definite matrix A

$$A = LL^*,$$

where L is a lower triangular matrix with positive diagonal entries. The Cholesky decomposition is unique, it is also applicable for complex Hermitian positive definite matrices.

4. QR decomposition: applicable to $m \times n$ matrix A

$$A = QR,$$

where Q is an orthogonal matrix of size $m \times m$, and R is an upper triangular matrix of size $m \times n$. For details of QR decomposition, please refer to Section 4.6.

5.1.2 Decompositions Based on Eigenvalues

1. Eigen-decomposition, also called spectral decomposition: applicable to square matrix A with distinct eigenvectors (not necessarily distinct eigenvalues).

$$A = Q\Lambda Q^{-1},$$

where Λ is a diagonal matrix formed from the eigenvalues of A, and the columns of Q are the corresponding eigenvectors of A. For details, see Section 3.2.

2. Jordan decomposition, also called Jordan normal form, or Jordan canonical form, applicable to a square matrix A. Jordan normal form generalizes the eigen-decomposition to cases where there are repeated eigenvalues and cannot be diagonalized. For details, see Section 3.3.

3. Spectral decomposition for a real, symmetric matrix A.

$$A = Q\Lambda Q^T,$$

where Λ is a diagonal matrix formed from the eigenvalues of A, and Q is an orthogonal matrix, whose columns are the corresponding orthonormal eigenvectors of A. For details, see Section 4.5.

4. Singular value decomposition: applicable to $m \times n$ matrix A.

$$A = U\Sigma V^*,$$

where Σ is an $m \times n$ matrix where the only non-zero entries are the non-negative diagonal entries in non-increasing order, U and V are unitary matrices, and V^* denotes the conjugate transpose of V (or simply the transpose, if V contains real numbers only). The diagonal elements of Σ are called the singular values of A. For details of SVD, please refer to Section 4.7.

5.1.3 Examples

LU Decomposition

In many applications involving solving systems of linear equations, one needs to solve $A\mathbf{x} = \mathbf{b}$ for many different vectors $\mathbf{b}$ with a fixed nonsingular matrix A. Gaussian elimination with pivoting is the most efficient and accurate way to solve a linear system. The key component of this method is to decompose A itself.

We observe if a system is in the following form

$$\begin{bmatrix} a_{11} & 0 & 0 & \cdots & 0 \\ 0 & a_{22} & 0 & \cdots & 0 \\ \vdots & \vdots & \vdots & \vdots & \vdots \\ 0 & 0 & 0 & \vdots & a_{nn} \end{bmatrix} \begin{bmatrix} x_1 \\ x_2 \\ \vdots \\ x_n \end{bmatrix} = \begin{bmatrix} b_1 \\ b_2 \\ \vdots \\ b_n \end{bmatrix}$$

then the system has a unique solution that is easy to obtain

$$x_i = \frac{b_i}{a_{ii}}. \quad i = 1, \ldots, n$$

On the other hand, if the system is given as a lower triangular matrix as

$$\begin{bmatrix} a_{11} & 0 & 0 & \cdots & 0 \\ a_{21} & a_{22} & 0 & \cdots & 0 \\ \vdots & \vdots & \vdots & \vdots & \vdots \\ a_{31} & a_{32} & a_{33} & \vdots & a_{nn} \end{bmatrix} \begin{bmatrix} x_1 \\ x_2 \\ \vdots \\ x_n \end{bmatrix} = \begin{bmatrix} b_1 \\ b_2 \\ \vdots \\ b_n \end{bmatrix};$$

then the system can be solved via forward substitution and has a unique solution

$$x_i = \frac{b_i - \sum_{j=1}^{i-1} a_{ij} x_j}{a_{ii}}. \quad i = 1, \ldots, n$$

Also, if the matrix is given in an upper triangular form

$$\begin{bmatrix} a_{11} & a_{12} & a_{13} & \cdots & a_{1n} \\ 0 & a_{22} & a_{23} & \cdots & a_{2n} \\ \vdots & \vdots & \vdots & \vdots & \vdots \\ 0 & 0 & 0 & \vdots & a_{nn} \end{bmatrix} \begin{bmatrix} x_1 \\ x_2 \\ \vdots \\ x_n \end{bmatrix} = \begin{bmatrix} b_1 \\ b_2 \\ \vdots \\ b_n \end{bmatrix};$$

then the system can be solved via backward substitution and has a unique solution

$$x_i = \frac{b_i - \sum_{j=i+1}^{n} a_{ij} x_j}{a_{ii}}. \quad i = n, \ldots, 1$$

LU decomposition, or LU factorization, stands for "lower upper," which factors a matrix as the product of a lower triangular matrix and an upper triangular matrix. The LU decomposition can be viewed as the matrix form of Gaussian elimination. Let A be an $n \times n$ nonsingular square matrix. One can perform row operation on A via a sequence of elementary matrices $E_1, \ldots, Ek$ to obtain an upper triangular matrix

$$U = E_k \ldots E_1 A.$$

Hence,

$$A = E_1^{-1} \cdots E_k^{-1} U = (E_1^{-1} \cdots E_k^{-1}) U = LU, \quad \text{where} \quad L = E_1^{-1} \cdots E_k^{-1}.$$

Since each E_i is a lower triangular matrix, the product of the inverse of elementary matrices is a lower triangular matrix. Thus A can be factored as

$$A = LU = \begin{bmatrix} l_{11} & 0 & 0 & \cdots & 0 \\ l_{21} & l_{22} & 0 & \cdots & 0 \\ \vdots & \vdots & \vdots & \vdots & \vdots \\ l_{31} & l_{32} & l_{33} & \vdots & l_{nn} \end{bmatrix} \begin{bmatrix} u_{11} & u_{12} & u_{13} & \cdots & u_{1n} \\ 0 & u_{22} & u_{23} & \cdots & u_{2n} \\ \vdots & \vdots & \vdots & \vdots & \vdots \\ 0 & 0 & 0 & \cdots & u_{nn} \end{bmatrix}$$

Furthermore, if A is a nonsingular matrix, then for each L, the upper triangular matrix U is unique but an LU decomposition is not unique. There can be more than one such LU decomposition for a matrix A. For example

$$\begin{bmatrix} 4 & 3 \\ 6 & 3 \end{bmatrix} = \begin{bmatrix} 1 & 0 \\ 1.5 & 1 \end{bmatrix} \begin{bmatrix} 4 & 3 \\ 0 & -3/2 \end{bmatrix}, \quad \text{or} \quad \begin{bmatrix} 4 & 3 \\ 6 & 3 \end{bmatrix} = \begin{bmatrix} 4 & 0 \\ 6 & 1 \end{bmatrix} \begin{bmatrix} 1 & 3/4 \\ 0 & -3/2 \end{bmatrix}$$

This is because when we compute LU decomposition by elementary transformations, the matrix U depends on the matrix L. To find out the unique LU decomposition, it is necessary to put some restriction on L and U matrices. For example, we can require all the entries of its main diagonal to be ones.

Therefore, given a system of linear equations in a matrix form

$$A\mathbf{x} = \mathbf{b},$$

to solve the equation for $\mathbf{x}$ given A and $\mathbf{b}$, we can decompose $A = LU$, then the solution is done in two logical steps: first, solve the equation $Ly = \mathbf{b}$ for y via forward substitution; then solve the equation $U\mathbf{x} = \mathbf{y}$ for $\mathbf{x}$ via backward substitution.

Remark 5.1.1: Not all non-singular matrices have an LU factorization. It is a proven result [HJ85] that: A matrix $A = (a_{ij}) 1 \leq i, j \leq n$ has an LU factorization if and only if the leading principal submatrices A_k of A are nonsingular for $k = 1, \ldots, n$.

The leading principal submatrix $A_k = (a_{ij})1 \le i, j \le k$ consists of the entries from the first k rows and k columns of A.

Rank Decomposition

Given an $m \times n$ matrix A of rank r, a rank decomposition or rank factorization of A is a product $A = CF$, where C is an $m \times r$ matrix and F is an $r \times n$ matrix.

Every finite-dimensional matrix has a rank decomposition. Since the rank$A = r$, there are r linearly independent columns in A; equivalently, the dimension of the column space of A is r. Thus, let $\mathbf{c}_1, \ldots, \mathbf{c}_r$ be any basis for the column space of A and let $C = [\mathbf{c}_1, \ldots, \mathbf{c}_r]$. Therefore, every column vector of A is a linear combination of the columns of C. That is, if

$$A = [\mathbf{a}_1, \ldots, \mathbf{a}_n], \quad \mathbf{a}_i \in \mathbb{R}^m$$

then the j-th column of A can be expressed as

$$\mathbf{a}_j = f_{1j}\mathbf{c}_1 + f_{2j}\mathbf{c}_2 + \cdots + f_{rj}\mathbf{c}_r,$$

where f_{ij} are the scalar coefficients of $\mathbf{a}_j$ in terms of the basis $\mathbf{c}_1, \ldots, \mathbf{c}_r$. This implies that $A = CF$, where $F = \{f_{ij}\}_{i=1,j=1}^{i=m,j=r}$ is an $r \times n$ matrix. For example,

$$A = \begin{bmatrix} 1 & 3 & 1 & 4 \\ 2 & 7 & 3 & 9 \\ 1 & 5 & 3 & 1 \\ 1 & 2 & 0 & 8 \end{bmatrix} \text{ with reduced echelon form} \begin{bmatrix} 1 & 0 & -2 & 0 \\ 0 & 1 & 1 & 0 \\ 0 & 0 & 0 & 1 \\ 0 & 0 & 0 & 0 \end{bmatrix}$$

Then C is obtained by removing the third column of A, and F is formed by removing the last row of zeroes in the reduced echelon form. Thus

$$A = CF, \quad \text{where} \quad C = \begin{bmatrix} 1 & 3 & 4 \\ 2 & 7 & 9 \\ 1 & 5 & 1 \\ 1 & 2 & 8 \end{bmatrix}, \quad F = \begin{bmatrix} 1 & 0 & -2 & 0 \\ 0 & 1 & 1 & 0 \\ 0 & 0 & 0 & 1 \end{bmatrix}$$

We note that the rank decomposition is not unique. If $A = C_1F_1$ is a rank factorization, then $C_2 = C_1R$ and $F_2 = R^{-1}F_1$ gives another rank factorization for any invertible $r \times r$ matrix R.

Cholesky Decomposition

In linear algebra, the Cholesky decomposition or Cholesky factorization is a decomposition of a Hermitian, positive-definite matrix A into the product $A = LL^*$ of a lower triangular matrix L with real and positive diagonal entries and its conjugate transpose L^*. If A has real entries, L has real entries as well, and $A = LL^T$.

Every Hermitian positive-definite matrix (and thus also every real-valued symmetric positive-definite matrix) has a unique Cholesky decomposition, that is, there is only one lower triangular matrix L with strictly positive diagonal entries such that $A = LL^*$. If the matrix A is Hermitian and positive semi-definite, then it still has a decomposition of the form $A = LL^*$ where the diagonal entries of L are allowed to be zero. However, the decomposition need not be unique when A is positive semi-definite.

The algorithm to calculate the matrix L for the decomposition is a modified version of Gaussian elimination. The recursive algorithm starts with $i := 1$ and

$$A_1 := A.$$

At step i, after performing Gaussian elimination to matrix A_1, the matrix A_i has the following form:

$$A_i = \begin{pmatrix} I_{i-1} & \mathbf{0} & \mathbf{0} \\ \mathbf{0} & a_{ii} & \mathbf{b}_i^* \\ \mathbf{0} & \mathbf{b}_i & B_i \end{pmatrix}$$

where $\mathbb{I}i_{-1}$ denotes the identity matrix of dimension $i - 1$. Define the matrix L_i by

$$L_i := \begin{pmatrix} \mathbb{I}_{i-1} & \mathbf{0} & \mathbf{0} \\ \mathbf{0} & \sqrt{a_{ii}} & \mathbf{0} \\ \mathbf{0} & \frac{1}{\sqrt{a_{ii}}}\mathbf{b}_i & \mathbb{I}_{i-1} \end{pmatrix}$$

then

$$A_i = L_i A_{i+1} L_i^*$$

where

$$A_{i+1} = \begin{pmatrix} I_{i-1} & \mathbf{0} & \mathbf{0} \\ \mathbf{0} & 1 & \mathbf{0} \\ \mathbf{0} & \mathbf{0} & B_i - \frac{1}{a_{ii}}\mathbf{b}_i\mathbf{b}_i^* \end{pmatrix}, \mathbf{b}_i\mathbf{b}_i^* = \begin{bmatrix} b_{i1} \\ \vdots \\ b_{i(n-i)} \end{bmatrix} \begin{bmatrix} \overline{b}_{i1} \cdots \overline{b}_{i(n-i)} \end{bmatrix}$$

Repeat this for $i = 1, \ldots, n$, and $A_{n+1} = \mathbb{I}$. Hence, the lower triangular matrix

$$L := L_1 L_2 \ldots L_n$$

Mathematica [Wol15], Maple [Map16], MATLAB [TM], Sage [SJ05], and many other computer software packages can compute matrix decompositions. MATLAB is a popular multi-paradigm numerical computing environment developed by Math Works. MATLAB allows matrix manipulations, plotting of functions and data, implementation of algorithms, creation of user interfaces, and interfacing with programs written in other languages, including C, C++, C#, Java, Fortran, and Python. MATLAB is a power tool for both students and engineers. Below, we use MATLAB to create a Pascal's matrix, then perform LU decomposition, QR decomposition, and Cholesky decomposition for this Pascal's matrix.

```
>>X=pascal(4) %generate a Pascal's matrix
X =
1  1   1   1
1  2   3   4
1  3   6  10
1  4  10  20
>> [L,U]=lu(X) %LU decomposition
L =
1.0000        0        0        0
1.0000   0.3333   1.0000        0
1.0000   0.6667   1.0000   1.0000
1.0000   1.0000        0        0
U =
1.0000   1.0000   1.0000   1.0000
     0   3.0000   9.0000  19.0000
     0        0  -1.0000  -3.3333
>> [Q,R]=qr(X) %QR decomposition
Q =
-0.5000   0.6708    0.5000    0.2236
-0.5000   0.2236   -0.5000   -0.6708
-0.5000  -0.2236   -0.5000    0.6708
-0.5000  -0.6708    0.5000   -0.2236
R =
-2.0000  -5.0000  -10.0000  -17.5000
      0  -2.2361   -6.7082  -14.0872
      0        0    1.0000    3.5000
      0        0         0   -0.2236
>> R=chol(X) %Cholesky decomposition, where R=L^*
R =
```

```
1  1  1  1
0  1  2  3
0  0  1  3
0  0  0  1
```

Since we investigated decomposition associated with the eigenvalues in previous chapters, below we only provide a few examples.

Eigen-decomposition

EXAMPLE 5.1.1

Taking a matrix $A = \begin{bmatrix} 1 & 0 \\ 1 & 3 \end{bmatrix}$. Since

$$\det(A - \lambda\mathbb{I}) = (1 - \lambda)(3 - \lambda) = 0.$$

There are two eigenvalues $\lambda_1 = 1$ and $\lambda_2 = 3$.

If $\lambda = 1$, then $A\mathbf{v} = 1\mathbf{v}$ gives $\mathbf{v} = \begin{bmatrix} -2 \\ 1 \end{bmatrix}$.

If $\lambda = 3$, then $A\mathbf{v} = 3\mathbf{v}$ gives $\mathbf{v} = \begin{bmatrix} 0 \\ 1 \end{bmatrix}$.

Thus the eigen-decomposition is

$$A = \begin{bmatrix} 1 & 0 \\ 1 & 3 \end{bmatrix} = \begin{bmatrix} -2 & 0 \\ 1 & 1 \end{bmatrix} A = \begin{bmatrix} 1 & 0 \\ 0 & 3 \end{bmatrix} \begin{bmatrix} -2 & 0 \\ 1 & 1 \end{bmatrix}^{-1} = Q\Lambda Q^{-1}$$

Jordan Canonical Decomposition

EXAMPLE 5.1.2

Suppose an $n \times n$ matrix A is of the form

$$A = \begin{bmatrix} 5 & 4 & 2 & 1 \\ 0 & 1 & -1 & -1 \\ -1 & -1 & 3 & 0 \\ 1 & 1 & -1 & 2 \end{bmatrix}.$$

Check that

$$\det(A - \lambda\mathbb{I}) = 0 \quad \Rightarrow \quad \lambda = 1, 2, 4, 4.$$

We check that the dimension of the eigenspace corresponding to the eigenvalue 4 is 1, so A is not diagonalizable. However, there is an invertible matrix P such that $A = P^{-1}J\,P$, where

$$J = \begin{bmatrix} 2 & 0 & 0 & \\ 0 & 1 & 0 & 0 \\ 0 & 0 & 4 & 1 \\ 0 & 0 & 0 & 4 \end{bmatrix}.$$

where J is the Jordan normal form of A.

Of course, we can hand compute matrices J and P for Example 5.1.2, but many computer software packages have built in code to complete this task. The following is a Sage [SJ05] code for the computation of Jordan canonical form in Example 5.1.2.

```
A=matrix(QQ, [[5, 4, 2, 1],[0,1,-1,-1], [-1,-1,3,0],[1,1,-1,2]])
A.jordan_form(transformation=True)
([2|0|0 0]
[-+-+---]
[0|1|0 0]
[-+-+---]
[0|0|4 1]
[0|0|0 4],
[ 1  1  1 1]
[-1 -1  0 0]
[ 0  0 -1 0]
[ 1  0  1 0])
```

Spectral Decomposition

EXAMPLE 5.1.3

Let $A = \begin{bmatrix} 3 & 2 \\ 2 & 3 \end{bmatrix}$. We continue with Sage code to obtain the minimal polynomial, eigenvalues, and eigenvectors.

```
B=matrix(QQ, [[3, 2],[2,3]])
B.minpoly()
x^2 - 6*x + 5
B.eigenvalues()
[5, 1]
B.eigenmatrix_left()
```

```
([5 0],[0 1], [ 1 1],[ 1 -1])
```

Hence the eigen-decomposition is

$$A = \begin{bmatrix} 1 & 1 \\ 1 & -1 \end{bmatrix} \begin{bmatrix} 5 & 0 \\ 0 & 1 \end{bmatrix} \begin{bmatrix} 1 & 1 \\ 1 & -1 \end{bmatrix}^{-1}.$$

Since A is a symmetric matrix, we can compute an orthonormal basis to obtain a spectral decomposition

$$A = \begin{bmatrix} 1/\sqrt{2} & 1/\sqrt{2} \\ 1/\sqrt{2} & -1/\sqrt{2} \end{bmatrix} \begin{bmatrix} 5 & 0 \\ 0 & 1 \end{bmatrix} \begin{bmatrix} 1/\sqrt{2} & 1/\sqrt{2} \\ 1/\sqrt{2} & -1/\sqrt{2} \end{bmatrix}^{-1}$$

5.2 Iterative Methods to Solve Linear Systems Numerically

Solving linear systems or matrix equations is a classical problem in matrix computations. In general, there are four methods to solve a linear system.

1. Use Cramer's rule to solve the system, if the matrix is a full-rank square matrix.
2. Compute the inverse or pseudoinverse of the system matrix.
3. Use matrix decomposition methods.
4. Use iterative methods.

Note that computing and applying the inverse matrix is much more expensive and often numerically less stable than applying one of the other algorithms. Hence the first two options are not used in matrix computations, especially for high-dimensional matrices.

The inverse of nonsingular square matrix A can be computed by the methods like Gauss-Jordan or LU decomposition. If A is not square, then A^TA and AA^T become square. Hence, it may be possible to apply Gauss-Jordan or LU decomposition to compute the inverse of A^TA or AA^T. Otherwise, a pseudoinverse (an inverse-like matrix) can be computed by singular value decomposition (SVD). Please see Section 5.1.1 for details of using ma- trix decompositions to solve linear systems.

In this section, we will focus on iterative methods to solve $A\mathbf{x} = \mathbf{b}$. A method of this type is a mathematical procedure that begins with an approximation to the solution, $\mathbf{x}_0$, then generates a sequence of improved approximations $\mathbf{x}_1, \ldots, \mathbf{x}_n$ that converge to the exact solution. This approach is appealing in engineering because it can be stopped as soon as the n-th approximation has an acceptable precision. Iterative methods are often useful for large and sparse system, but iterative methods can be unreliable; since for some problems they may have slow convergence, or they may not converge at all. There are two fundamental iterative methods: the Jacobi method and the GaussSeidel method.

5.2.1 Jacobi Iiterative Method

First, the Jacobi iterative method is an algorithm for determining the solutions of a diagonally dominant system of linear equations. A square matrix $A = (a_{ij})$ where a_{ij} denotes the entry in the i-th row and j-th column is said to be diagonally dominant if

$$|a_{ii}| \geq \sum_{j \neq 1} |a_{ij}| \text{ for all } i.$$

Let $A\mathbf{x} = \mathbf{b}$ be a square system of n linear equations, where:

$$A = \begin{bmatrix} a_{11} & a_{12} & \cdots & a_{1n} \\ a_{21} & a_{22} & \cdots & a_{2n} \\ \vdots & \vdots & \ddots & \vdots \\ a_{n1} & a_{n2} & & a_{nn} \end{bmatrix}, \quad \mathbf{x} = \begin{bmatrix} x_1 \\ x_2 \\ \vdots \\ x_n \end{bmatrix}, \quad \mathbf{b} = \begin{bmatrix} b_1 \\ b_2 \\ \vdots \\ b_n \end{bmatrix}$$

Then A can be decomposed into a diagonal component D, and the remainder R:

$$A = D + R \text{ where } D = \begin{bmatrix} a_{11} & 0 & \cdots & 0 \\ 0 & a_{22} & \cdots & 0 \\ \vdots & \vdots & \ddots & \vdots \\ 0 & 0 & \cdots & a_{nn} \end{bmatrix}, \text{and } R = \begin{bmatrix} 0 & a_{12} & \cdots & a_{1n} \\ a_{21} & 0 & \cdots & a_{2n} \\ \vdots & \vdots & \ddots & \vdots \\ a_{n1} & a_{n2} & \cdots & 0 \end{bmatrix}$$

We note that $R = L + U$ where L and U are the strictly lower and upper parts of A.

Then the solution is then obtained iteratively via

$$\begin{aligned} x^{(k+1)} &= D^{-1}(\mathbf{b} - R\mathbf{x}^{(k)}), \\ &= T\mathbf{x}^{(k)} + C, \text{ where } T = -D^{-1}R \text{ and } C = D^{-1}\mathbf{b}, \end{aligned}$$

where $\mathbf{x}^{(k)}$ is the k-th iteration of $\mathbf{x}$ and $\mathbf{x}^{(k+1)}$ is the $(k + 1)$-th iteration of $\mathbf{x}$. The the solution for the system is

$$x_i^{(k+1)} = \frac{1}{a_{ii}}\left(b_i = \sum_{j \neq 1} a_{ij} x_j^{(k)}\right), \quad i = 1, 2, \ldots, n.$$

We will use the following example to illustrate the method.

EXAMPLE 5.2.1

Let

$$A=\begin{bmatrix}2 & 1\\ 5 & 7\end{bmatrix},\quad \mathbf{b}=\begin{bmatrix}11\\ 13\end{bmatrix}\text{ and }\mathbf{x}^{(0)}=\begin{bmatrix}1\\ 1\end{bmatrix},$$

where $\mathbf{x}^{(0)}$ is the initial approximation. Then, the above procedure yields

$$\begin{aligned}\mathbf{x}^{(k+1)} &= D^{-1}(\mathbf{b}-R\mathbf{x}^{(k)})\\ &= T\mathbf{x}^{(k)}+C,\text{ where } T=-D^{-1}R \text{ and } C=D^{-1}\mathbf{b},\text{ and}\end{aligned}$$

$$D^{-1}=\begin{bmatrix}1/2 & 0\\ 0 & 1/7\end{bmatrix},\quad L=\begin{bmatrix}0 & 0\\ 5 & 0\end{bmatrix}\text{ and }U=\begin{bmatrix}0 & 1\\ 0 & 0\end{bmatrix},$$

$$T=\begin{bmatrix}1/2 & 0\\ 0 & 1/7\end{bmatrix}\left\{\begin{bmatrix}0 & 0\\ -5 & 0\end{bmatrix}+\begin{bmatrix}0 & -1\\ 0 & 0\end{bmatrix}\right\}=\begin{bmatrix}0 & -1/2\\ -5/2 & 0\end{bmatrix}$$

$$C=\begin{bmatrix}1/2 & 0\\ 0 & 1/7\end{bmatrix}\begin{bmatrix}11\\ 13\end{bmatrix}=\begin{bmatrix}11/2\\ 13/7\end{bmatrix}.$$

The iteration gives

$$\mathbf{x}^{(1)}=\begin{bmatrix}0 & -1/2\\ -5/7 & 0\end{bmatrix}\begin{bmatrix}1\\ 1\end{bmatrix}+\begin{bmatrix}11/2\\ 13/7\end{bmatrix}=\begin{bmatrix}5.0\\ 8/7\end{bmatrix}\approx\begin{bmatrix}5\\ 1.143\end{bmatrix}$$

$$\mathbf{x}^{(2)}=\begin{bmatrix}0 & -1/2\\ -5/7 & 0\end{bmatrix}\begin{bmatrix}5.0\\ 8/7\end{bmatrix}+\begin{bmatrix}11/2\\ 13/7\end{bmatrix}=\begin{bmatrix}69/14\\ -12/7\end{bmatrix}\approx\begin{bmatrix}4.929\\ -1.714\end{bmatrix}$$

$\vdots$ process repeats until an acceptable precision is achieved

$$\mathbf{x}^{(25)}=\begin{bmatrix}7.111\\ -3.222\end{bmatrix}.$$

5.2.2 Gauss-Seidel iterative method

The Gauss-Seidel iterative method is very similar to the Jacobi method. But unlike the Jacobi method, the Gauss-Seidel method includes successive displacement. That is, the computation of $\mathbf{x}^{(k+1)}$ uses only the elements of $x_i^{(k+1)}$ that have already been computed, and only the elements of $\mathbf{x}^{(k)}$ that have not yet to be advanced to the $(k+1)$-th iteration.

Let $A\mathbf{x}=\mathbf{b}$ be a square system of n linear equations, where:

$$A=\begin{bmatrix}a_{11} & a_{12} & \cdots & a_{1n}\\ a_{21} & a_{22} & \cdots & a_{2n}\\ \vdots & \vdots & \ddots & \vdots\\ a_{n1} & a_{n2} & & a_{nn}\end{bmatrix},\quad \mathbf{x}=\begin{bmatrix}x_1\\ x_2\\ \vdots\\ x_n\end{bmatrix},\quad \mathbf{b}=\begin{bmatrix}b_1\\ b_2\\ \vdots\\ b_n\end{bmatrix}$$

Then A can be decomposed into its lower triangular component and its strictly upper triangular component given by:

$$A = L_* + U \text{ where } L_* = \begin{bmatrix} a_{11} & 0 & \cdots & 0 \\ 0 & a_{22} & \cdots & 0 \\ \vdots & \vdots & \ddots & \vdots \\ 0 & 0 & \cdots & a_{nn} \end{bmatrix}, U = \begin{bmatrix} 0 & a_{12} & \cdots & a_{1n} \\ a_{21} & 0 & \cdots & a_{2n} \\ \vdots & \vdots & \ddots & \vdots \\ a_{n1} & a_{n2} & \cdots & 0 \end{bmatrix}$$

The system of linear equations $A\mathbf{x} = \mathbf{b}$ can be rewritten as:

$$L_*\mathbf{x} = \mathbf{b} - Ux.$$

The Gauss-Seidel method now solves the left-hand side of the above expression for $\mathbf{x}$, using previous value for $\mathbf{x}$ on the right hand side:

$$\mathbf{x}^{(k+1)} = L_*^{-1}(\mathbf{b} - U\mathbf{x}^{(k)}) = T\mathbf{x}^{(k)} + C$$

where $T = -L_*^{-1}U$ and $C = -L_*^{-1}\mathbf{b}$

Using the lower triangular matrix L^*, the elements of $x^{(k+1)}$ can be computed sequentially using forward substitution:

$$x_i^{(k+1)} = \frac{1}{a_{ii}}\left(b_i - \sum_{j=1}^{i-1} a_{ij}x_j^{(k+1)} - \sum_{j=1+1}^{n} a_{ij}x_j^{(k)}\right), i = 1, 2, \ldots, n$$

The procedure terminates when an acceptable precision is achieved.

We will use the following example to illustrate this method.

EXAMPLE 5.2.1

Let $A\mathbf{x} = \mathbf{b}$ be the following:

$$A = \begin{bmatrix} 16 & 3 \\ 7 & -11 \end{bmatrix}, \quad \mathbf{b} = \begin{bmatrix} 11 \\ 13 \end{bmatrix} \text{ and } \mathbf{x}^{(0)} = \begin{bmatrix} 1 \\ 1 \end{bmatrix},$$

where $\mathbf{x}^{(0)}$ is the initial approximation. Then

$$\mathbf{x}^{(k+1)} = L_*^{-1}(\mathbf{b} - U\mathbf{x}^{(k)} = T\mathbf{x}^{(k)} + C, \text{ where}$$

$$L_* = \begin{bmatrix} 16 & 0 \\ 0 & -11 \end{bmatrix} \quad U = \begin{bmatrix} 0 & 3 \\ 0 & 0 \end{bmatrix},$$

$$T = -L_*^{-1}U = \begin{bmatrix} 0.000 & -0.1875 \\ 0.000 & -0.1193 \end{bmatrix}$$

$$C = L_*^{-1}\mathbf{b} = \begin{bmatrix} 0.6875 \\ -0.7443 \end{bmatrix}$$

The iteration formula gives:

$$\mathbf{x}^{(1)} = \begin{bmatrix} 0.000 & -0.1875 \\ 0.000 & -0.1193 \end{bmatrix} \begin{bmatrix} 1.0 \\ 1.0 \end{bmatrix} + \begin{bmatrix} 0.6875 \\ -0.7443 \end{bmatrix} = \begin{bmatrix} 0.5000 \\ -0.8636 \end{bmatrix}$$

$$\mathbf{x}^{(2)} = \begin{bmatrix} 0.000 & -0.1875 \\ 0.000 & -0.1193 \end{bmatrix} \begin{bmatrix} 0.5000 \\ -0.8636 \end{bmatrix} + \begin{bmatrix} 0.6875 \\ -0.7443 \end{bmatrix} = \begin{bmatrix} 0.8494 \\ -0.6413 \end{bmatrix}$$

$$\mathbf{x}^{(3)} = \begin{bmatrix} 0.000 & -0.1875 \\ 0.000 & -0.1193 \end{bmatrix} \begin{bmatrix} 0.8494 \\ -0.6413 \end{bmatrix} + \begin{bmatrix} 0.6875 \\ -0.7443 \end{bmatrix} = \begin{bmatrix} 0.8077 \\ -0.6678 \end{bmatrix}$$

$$\mathbf{x}^{(4)} = \begin{bmatrix} 0.000 & -0.1875 \\ 0.000 & -0.1193 \end{bmatrix} \begin{bmatrix} 0.8077 \\ -0.6678 \end{bmatrix} + \begin{bmatrix} 0.6875 \\ -0.7443 \end{bmatrix} = \begin{bmatrix} 0.8127 \\ -0.6646 \end{bmatrix}$$

$$\mathbf{x}^{(5)} = \begin{bmatrix} 0.000 & -0.1875 \\ 0.000 & -0.1193 \end{bmatrix} \begin{bmatrix} 0.8127 \\ -0.6646 \end{bmatrix} + \begin{bmatrix} 0.6875 \\ -0.7443 \end{bmatrix} = \begin{bmatrix} 0.8121 \\ -0.6650 \end{bmatrix}$$

$$\mathbf{x}^{(6)} = \begin{bmatrix} 0.000 & -0.1875 \\ 0.000 & -0.1193 \end{bmatrix} \begin{bmatrix} 0.8121 \\ -0.6650 \end{bmatrix} + \begin{bmatrix} 0.6875 \\ -0.7443 \end{bmatrix} = \begin{bmatrix} 0.8122 \\ -0.6650 \end{bmatrix}$$

$$\mathbf{x}^{(7)} = \begin{bmatrix} 0.000 & -0.1875 \\ 0.000 & -0.1193 \end{bmatrix} \begin{bmatrix} 0.8122 \\ -0.6650 \end{bmatrix} + \begin{bmatrix} 0.6875 \\ -0.7443 \end{bmatrix} = \begin{bmatrix} 0.8122 \\ -0.6650 \end{bmatrix}$$

The algorithm converges to the exact solution, and we check that:

$$\mathbf{x} = A^{-1}\mathbf{b} \approx \begin{bmatrix} 0.8122 \\ -0.6650 \end{bmatrix}.$$

5.3 Matrix over Principle Ideal Domain

5.3.1 Matrix Decomposition over PID

In abstract algebra, a principal ideal domain, or PID, is an integral domain in which every ideal is principal, i.e., can be generated by a single element. More generally, a principal ideal ring is a non-zero commutative ring whose ideals are principal. The distinction is that a principal ideal ring may have zero divisors whereas a principal ideal domain cannot.

Principal ideal domains behave somewhat like the integers, with respect to divisibility: any element of a PID has a unique decomposition into prime elements; any two elements of a PID have a greatest common divisor. If gcd(x, y) = 1 for x and y in a PID, then every element of the PID can be written in the form $ax + by$ for some a, b in this PID.

Furthermore, one of the important properties of a PID is that if $(a_1) \subseteq \ldots \subseteq (a_2) \subseteq \ldots$is an increasing sequence of ideals, then there is an n such that $(a_n) = (a_{n+1}) = \ldots$. To see this, we note that $\cup(a_i)$ is also an ideal, and must be generated by an element b in a PID. Therefore, $\cup(a_i) = (b)$, and this forces $(a_n) = (a_{n+1}) = \ldots = (b)$.

In this section, we will discuss the structure theorem for finitely generated modules over a principal ideal domain from the point of view of matrices.

Let R be a principal ideal domain and let M be a finitely generated R-module. If $\{m_1, \ldots, m_n\}$ is a set of generators of M, then we have a surjective R-module homomorphism

$$\phi : R^n \rightarrow M, \quad (r_1, \ldots, r_n) \rightarrow \sum_{i=1}^{n} r_i m_i$$

Let $K = \ker \phi = \{(r_1, \ldots, r_n) \in R^n \mid \sum_{i=1}^{n} r_i m_i = 0\}$, then $M > R^n/K$. It is known that K is finitely generated. Suppose that $\{\mathbf{a}_1, \ldots, \mathbf{a}_m\} \subseteq R^n$ is a generating set for K, where $\mathbf{a}_i = (a_{i1}, a_{i2}, \ldots, a_{in})$, then we will refer to the matrix (a_{ij}) over R as the relation matrix for M relative to the generating set $\{m_1, \ldots, m_n\}$ of M and the generating set $\{\mathbf{a}_1, \ldots, \mathbf{a}_m\} \subseteq R^n$ of K. In the language of commutative algebra, K is call the **syzygy module** of the generating set $\{m_1, \ldots, m_n\}$ of M.

Lemma 5.3.1 Let M be a finitely generated R-module, with ordered generating set $\{m_1, \ldots, m_n\}$. Suppose that the submodule K is generated by $\{\mathbf{a}_1, \ldots, \mathbf{a}_m\}$. Let A be the $m \times n$ relation matrix relative to these generators.

1. Let $P \in M_{m \times m}(R)$ be an invertible matrix. If $[\mathbf{l}_1, \ldots, \mathbf{l}_m]$ are the rows of PA, then they generate K, and so PA is the relation matrix relative to $[m_1, \ldots, m_n]$ and $[\mathbf{l}_1, \ldots, \mathbf{l}_m]$.
2. Let $Q \in M_{n \times n}(R)$ be an invertible matrix and write $Q^{-1} = (q_{ij})$ If
$$t_j = \sum_i q_{ij} m_i, \text{ for } 1 \leq j \leq n,$$
then $[t_1, \ldots, t_n]$ is a generating set for M and the rows of AQ generate the corresponding relation submodule. Therefore, AQ is a relation matrix relative to $[t_1, \ldots, t_n]$.

3. Let P and Q be $m \times m$ and $n \times n$ invertible matrices, respectively. If $B = PAQ$, then B is the relation matrix relative to an appropriate ordered set of generators of M and of the corresponding relation submodule.

Proof:

1. Notice the matrix multiplication gives

$$\begin{bmatrix} \mathbf{l}_1 \\ \vdots \\ \mathbf{l}_m \end{bmatrix}\begin{bmatrix} m_1 \\ \vdots \\ m_n \end{bmatrix} = P_A \begin{bmatrix} m_1 \\ \vdots \\ m_n \end{bmatrix} = P\left(A \begin{bmatrix} m_1 \\ \vdots \\ m_n \end{bmatrix} \right) = P\mathbb{O} = \mathbb{O}$$

In addition that P is an invertible matrix, $A = P^{-1}[\mathbf{l}_1, \ldots, \mathrm{l}_m]^T$. Hence, the rows of the matrix A can be expressed in terms of $\{\mathbf{l}_1, \ldots, \mathbf{l}_m\}$. Matrix A can be considered as a relation matrix relative to the generating set $\{m_1, \ldots, m\} \subset M$ and $\{\mathbf{l}_1, \ldots, \mathbf{l}m\} \subset R^n$.

2. Notice the matrix multiplication gives

$$\begin{bmatrix} t_1 \\ \vdots \\ t_n \end{bmatrix} = Q^{-1}\begin{bmatrix} m_1 \\ \vdots \\ m_n \end{bmatrix} \Leftrightarrow \begin{bmatrix} m_1 \\ \vdots \\ m_n \end{bmatrix} = Q\begin{bmatrix} t_1 \\ \vdots \\ t_n \end{bmatrix}$$

Thus, $\{t_1, \ldots, t_n\}$ is a generating set for M. In addition,

$$\mathbb{O} = A\begin{bmatrix} m_1 \\ \vdots \\ m_n \end{bmatrix} = A\left(Q\begin{bmatrix} t_1 \\ \vdots \\ t_n \end{bmatrix} \right) = (AQ)\begin{bmatrix} m_1 \\ \vdots \\ m_n \end{bmatrix}$$

yields that the rows of AQ generate the corresponding relation submodule. Therefore, AQ is a relation matrix relative to $\{t_1, \ldots, t_n\}$.

3. The claim follows directly from (1) and (2).

Proposition 5.3.1 Suppose that A is a relation matrix for an R-module M. If there are invertible matrices P and Q for which

$$PAQ = \begin{bmatrix} a_1 & 0 & \cdots & 0 \\ 0 & a_2 & \cdots & 0 \\ 0 & 0 & \ddots & 0 \\ 0 & 0 & \cdots & a_n \end{bmatrix}$$

is a diagonal matrix, then $M \cong \oplus_{i=1}^{n} nR/(a_i)$.

Proof: The matrix PAQ above is the relation matrix for an ordered generating set $[m_1, \ldots, m_n]$ relative to a relation submodule generated by the rows of PAQ. If $f: R^n \to M$ is a homomorphism such that $f(r_1, \ldots, r_n) = \sum_{i=1}^{n} r_i m_i$, then $K = \ker(f)$. Thus, $M \cong R^n/K$. Since K is also the kernel of the surjective R-module homomorphism $g: R^n \to \oplus_{i=1}^{n} R/(a_i)$ such that $g(r_1, \ldots, r_n) = (r_1 + (a_1), \ldots, r_n + (a_n))$. Thus, $M \cong R^n/K \cong \oplus_{i=1}^{n} R/(a_i)$.

Definition 5.3.1

Let R be a principal ideal domain and let A be a $p \times n$ matrix with entries in R. We say that A is in **Smith normal form** if there are non-zero $a_1, \ldots, a_m \in R$ such that $a_i \mid a_{i+1}$ for each $i < m$, and

$$A = \begin{bmatrix} a_1 & & & & & & \\ & a_2 & & & & & \\ & & \ddots & & & & \\ & & & a_m & & & \\ & & & & 0 & & \\ & & & & & \ddots & \\ & & & & & & 0 \end{bmatrix}.$$

Theorem 5.3.1

If A is a matrix with entries in a principal ideal domain R, then there are invertible matrices P and Q over R such that PAQ is in Smith normal form.

Proof: We will only illustrate the argument for the case of 2 × 2 matrix. Consider $\begin{bmatrix} a & b \\ c & d \end{bmatrix}$, and let $e_0 = \gcd(a, c)$, and hence $e_0 = ax + cy$ for some $x, y \in R$. Thus

$$\begin{bmatrix} a & b \\ c & d \end{bmatrix} = \begin{bmatrix} e_0 a' & b \\ e_0 c' & d \end{bmatrix}, \text{ where } a'x + c'y = 1.$$

since

$$\begin{bmatrix} x & y \\ -a' & c' \end{bmatrix}^{-1} = \begin{bmatrix} c' & -y \\ a' & x \end{bmatrix},$$

and

$$\begin{bmatrix} x & y \\ -a' & c' \end{bmatrix}^{-1} \begin{bmatrix} a & b \\ c & d \end{bmatrix} = \begin{bmatrix} e_0 & bx + dy \\ -aa' + cc' & -ba' + dc' \end{bmatrix} \xrightarrow[\text{since } e_0 \mid (-aa' + cc')]{\text{row operaiton}} \begin{bmatrix} e_0 & u \\ 0 & v \end{bmatrix}.$$

An similar treatment applied to the first row via multiplying on the right by an invertible matrix and obtain a matrix to the form $\begin{bmatrix} e_1 & 0 \\ * & * \end{bmatrix}$ where $e_1 = \gcd(e_0, u)$.

Continuing this process, alternating between the first row and the first column, will produce a sequence of elements $e_0, e_1, \ldots$ such that $e_{i+1} \mid e_i$, $i \geq 0$. As ideals, this says $(e_0) \subseteq (e_1) \subseteq \ldots \subseteq (e_n) = \ldots = (e)$, since increasing sequence of principal ideals stabilizes in PID. Thus, in finitely many steps multiplying by invertible matrices P_i and Q_i, we derive

$$P_n \cdots P_1 \begin{bmatrix} a & b \\ c & d \end{bmatrix} Q_1 \cdots Q_n = \begin{cases} \begin{bmatrix} e & 0 \\ C & D \end{bmatrix} \\ \text{or} \\ \begin{bmatrix} e & C \\ 0 & D \end{bmatrix} \end{cases}, \text{ where } e \mid C.$$

Thus, one more row or column operation, we have a diagonal matrix $\begin{bmatrix} e & 0 \\ 0 & D \end{bmatrix}$.

Now, let $g = \gcd(e, D)$, then $e = ge'$ and $D = gD'$ and there exists m, $n \in R$ such that $g = em + Dn$, hence,

$$\begin{bmatrix} e & 0 \\ 0 & D \end{bmatrix} \to \begin{bmatrix} e & 0 \\ em & D \end{bmatrix} \to \begin{bmatrix} e & 0 \\ g = em + Dn & D \end{bmatrix} \to \begin{bmatrix} 0 & -e'D \\ g & D \end{bmatrix} \to \begin{bmatrix} g & 0 \\ 0 & -ge'D' \end{bmatrix}.$$

Thus,

$$P \begin{bmatrix} a & b \\ c & d \end{bmatrix} Q = \begin{bmatrix} g & 0 \\ 0 & -ge'D' \end{bmatrix},$$

and we conclude that there are invertible matrices P and Q over R such that PAQ is in Smith normal form.

A consequence of the existence of a Smith normal form is the following structure theorem of finitely generated modules over PID.

Corollary 5.3.1: Suppose M is a finitely generated module over a PID R, then there are elements $a_1, \ldots, a_m \in R$ such that $a_i \mid a_{i+1}$, $i = 1, \ldots, m - 1$, and $t \in \mathbb{Z}_{\geq 0}$ such that $M \cong R/(a_1) \oplus \cdots \oplus R/(a_m) \oplus R^t$.

Proof: Let A be a relation matrix for M, and let B be its Smith normal form, i.e., for some invertible matrices P, Q,

$$PAQ = B = \begin{bmatrix} a_1 & & & & & & \\ & a_2 & & & & & \\ & & \ddots & & & & \\ & & & a_m & & & \\ & & & & 0 & & \\ & & & & & \ddots & \\ & & & & & & 0 \end{bmatrix}.$$

Proposition 5.3.1 implies

$$\begin{aligned} M &\cong R/(a_1)\oplus\cdots\oplus R/(a_m)\oplus R/(0)\oplus\cdots\oplus R/(0) \\ &\cong R/(a_1)\oplus\cdots\oplus R/(a_m)\oplus Rt, \text{for some } t\in\mathbb{Z}_{\geq 0} \end{aligned}$$

Hence, we proved our claim.

It is known that a polynomial ring in x over the field $\mathbb{F}$, $\mathbb{F}[x]$ is a PID, and in the remainder of this section, we will focus on the case when $R = \mathbb{F}[x]$. Let $A \in M_n(\mathbb{F})$ be an $n \times n$ matrix, then we can consider $\mathbb{F}^n$ as a finitely generated $\mathbb{F}[x]$-module via

$$\mathbb{F}[x]\times\mathbb{F}^n \to \mathbb{F}^n : f(x)\mathbf{m} = f(A)\mathbf{m}.$$

Let $\mathbf{e}_1, \ldots, \mathbf{e}_n$ be the standard basis for $\mathbb{F}^n$. Now, we consider a $\mathbb{F}[x]$-module homomorphism:

$$\phi : \mathbb{F}[x]^n \to \mathbb{F}^n : \quad \phi(f_1(x),\ldots,f_n(x)) = \sum_{i=1}^{n} f_i(x)\mathbf{e}_i = \sum_{i=1}^{n} f_i(A)\mathbf{e}_i$$

Note, if $A = (a_{ij})$, then

$$\phi(x,x,\ldots,x) = \sum_{i=1}^{n} x\mathbf{e}_i = \sum_{i=1}^{n} A\mathbf{e}_i = \sum_{i=1}^{n} \begin{bmatrix} a_{1i} \\ \vdots \\ a_{ni} \end{bmatrix} = \sum_{i=1}^{n} a_{1_i}\mathbf{e}_1 + \cdots + \sum_{i=1}^{n} a_{ni}\mathbf{e}_n$$

thus

$$\sum_{i=1}^{n} x\mathbf{e}_i = \left(\sum_{i=1}^{n} a_{1_i}\mathbf{e}_1 + \cdots + \sum_{i=1}^{n} a_{ni}\mathbf{e}_n\right) = (x\mathbb{I} - A)\begin{bmatrix} 1 \\ \vdots \\ 1 \end{bmatrix} = 0.$$

Hence, $x\mathbb{I} - A$ is a relation matrix relative to $[\mathbf{e}_1, \ldots, \mathbf{e}_n]$.

From now on, let $\mathbf{v}_1, \ldots, \mathbf{v}_n$ be the rows of $x\mathbb{I} - A^T$, and let $\mathbf{E}_1, \ldots, \mathbf{E}_n$ be the standard basis vectors of $\mathbb{F}[x]^n$.

Lemma 5.3.2: Let $\sum_{i=1}^{n} f_i(x)\mathbf{E}_i \in \mathbb{F}[x]^n$. Then there are $g_i(x) \in \mathbb{F}[x]$ and $b_i \in \mathbb{F}$ such that

$$\sum_{i=1}^{n} f_i(x)\mathbf{E}_i = \sum_{i=1}^{n} g_i(x)\mathbf{v}_i + \sum_{i=1}^{n} b_i\mathbf{E}_i.$$

Proof: We will prove the claim by induction on $\max(\deg(f_i(x)))_{i=1}^{n}$.

If $m = 0$, then $f_i(x)$ are constant, hence let $b_i = f_i(x)$, and the claim is true.

Assume that for $m > 0$, and the claim is true for polynomial vectors $[f_1(x), \ldots, f_n(x)]$ whose maximal degree is no more than m. Hence, for $i = 1, \ldots, n$. by division algorithm, $f_i(x) = q_i(x)(x - a_{ii}) + r_i$, where $\deg(q_i(x)) \le \deg(f_i(x)) - 1$, $r_i \in \mathbb{F}$. Consider

$$\begin{aligned}
[f_1(x), 0, \ldots, 0] &= [q_1(x)(x - a_{11}) + r_1, 0, \ldots, 0] \\
&= q_1(x)[(x - a_{11}), 0, \ldots, 0] + [r_1, 0, \ldots, 0] \\
&= q_1(x)[(x - a_{11}), -a_{21}, \ldots, -a_{n1}] + [r_1, q_1(x)a_{21}, \ldots, q_1(x)a_{n1}] \\
&= q_1(x)\mathbf{v}_1 + [r_1, q_1(x)a_{21}, \ldots, q_1(x)a_{n1}].
\end{aligned}$$

Notice, that the entries of the second polynomial vector has degree strictly less than $\deg(f_1(x))$. Repeating this process for each $f_i(x)$, we have

$$\begin{aligned}
\sum_{i=1}^{n} f_i(x)\mathbf{E}_i &= \sum_{i=1}^{n} q_i(x)\mathbf{v}_i + \sum_{i=1}^{n} h_i(x)\mathbf{E}_i, \quad \deg(h_i(x)) < \deg(f_i(x)) < m \\
&= \sum_{i=1}^{n} q_i(x)\mathbf{v}_i + \left(\sum_{i=1}^{n} h_i(x)\mathbf{v}_i + \sum_{i=1}^{n} b_i\mathbf{E}_i\right) \text{ by induction hypothesis} \\
&= \sum_{i=1}^{n} (q_i(x) + h_i(x))\mathbf{v}_i + \sum_{i=1}^{n} b_i\mathbf{E}_i = \sum_{i=1}^{n} g_i(x)\mathbf{v}_i + \sum_{i=1}^{n} b_i\mathbf{E}_i
\end{aligned}$$

Thus, the claim is true by induction.

Proposition 5.3.2. If ϕ is a $\mathbb{F}[x]$-module homomorphism defined by

$$\phi : \mathbb{F}[x]^n \to \mathbb{F}^n : \phi(f(x), \ldots, f_n(x)) = \sum_{i=1}^{n} f_i(x)\mathbf{e}_i,$$

then $\ker(\phi)$ is generated by the rows of $x\mathbb{I} - A^T$.

Proof: Let M be the submodule of $\mathbb{F}[x]^n$ generated by $\mathbf{v}_1, \ldots, \mathbf{v}_n$, the rows of $x\mathbb{I} - A^T$. To show $M = \ker(\phi)$, we first note that $\mathbf{v}_i \in \ker(\phi)$, thus $M \subseteq \ker(\phi)$. Thus, we only need to show that $\ker(\phi) \subseteq \mathrm{M}$.

Let $\sum_{i=1}^{n} f_i(x)\mathbf{E}_i \in \ker(\phi)$, and by Lemma 5.3.2,

$$\sum_{i=1}^{n}(f_i(x)\mathbf{E}_i = \sum_{i=1}^{n} g_i(x)\mathbf{v}_i + \sum_{i=1}^{n} b_i\mathbf{E}_i, g_i(x) \in F[x],\ b_i \in \mathbb{F}.$$

Since $\mathbf{v}_i \in \ker(\phi)$, we must have that $\sum_{i=1}^{n} b_i\mathbf{E}_i \in \ker(\phi)$. Moreover,

$$\phi\left(\sum_{i=1}^{n} b_i\mathbf{E}_i\right) = \sum_{i=1}^{n} b_i\mathbf{e}_i \in \mathbb{F}^n \quad \Rightarrow \quad b_i = 0, \quad \forall i = 1,\ldots,n.$$

Thus,

$$\sum_{i=1}^{n} f_i(x)\mathbf{E}_i = \sum_{i=1}^{n} g_i(x)\mathbf{v}_i \in M.$$

Thus, $\ker(\phi) \subseteq M$.

Therefore, we conclude $\ker(\phi) = M$.

Since $\mathbb{F}[x]/(1)$ is the zero module, with the above notation, we can conclude that

Corollary 5.3.2: If

$$B = \begin{bmatrix} 1 & & & & & \\ & \vdots & & & & \\ & & 1 & & & \\ & & & f_1(x) & & \\ & & & & \vdots & \\ & & & & & f_m(x) \end{bmatrix}$$

is the Smith normal form of A, then as a $\mathbb{F}[x]$-module,

$$\mathbb{F}^n \cong \mathbb{F}[x]/(f_1(x)) \oplus \cdots \oplus \mathbb{F}[x]/(fm(x))$$

The invariant factors of $\mathbb{F}^n$ are $f_1(x), \ldots, f_m(x)$.

It is difficult to hand compute Smith normal form. Many computer algebra systems have build-in code to compute Smith normal form. Below, we present one simple example via Maple. In this example, we first create a matrix Ax over the ring $\mathbb{R}[x]$, then, we compute the Smith normal form of matrix Ax.

```
>Ax:= Matrix([[1,2*x,2*x^2+2*x],[1,6*x,6*x^2+6*x],[1,3,x]]);
```

$$Ax := \begin{bmatrix} 1 & 2x & 2x^2+2x \\ 1 & 6x & 6x^2+6x \\ 1 & 3 & x \end{bmatrix}$$

```
>S := SmithForm(Ax);
```

$$S := \begin{bmatrix} 1 & 0 & 0 \\ 0 & 1 & 0 \\ 0 & 0 & x^2+3/2x \end{bmatrix}$$

5.4 Exercises

1. Find LU decompositions for the matrices

$$A = \begin{bmatrix} 3 & 4 \\ 1 & 2 \end{bmatrix}, \quad B = \begin{bmatrix} 1 & 4 & 7 \\ 2 & 5 & 8 \\ 3 & 6 & 9 \end{bmatrix}$$

2. Verify the following Cholesky decomposition

$$\begin{bmatrix} 25 & 15 & 5 \\ 15 & 18 & 0 \\ -5 & 0 & 11 \end{bmatrix} = \begin{bmatrix} 5 & 0 & 0 \\ 3 & 3 & 0 \\ -1 & 1 & 3 \end{bmatrix} \begin{bmatrix} 5 & 3 & -1 \\ 0 & 3 & 1 \\ 0 & 0 & 3 \end{bmatrix}$$

3. Verify the following Cholesky decomposition

$$\begin{bmatrix} 25 & -50 \\ -50 & 101 \end{bmatrix} = \begin{bmatrix} 5 & 0 \\ -10 & 1 \end{bmatrix} \begin{bmatrix} 5 & -10 \\ 0 & 1 \end{bmatrix}$$

4. Verify the following QR decomposition

$$\begin{bmatrix} 3 & -6 \\ 4 & -8 \\ 0 & 1 \end{bmatrix} = \begin{bmatrix} 3/5 & 0 \\ 4/5 & 0 \\ 0 & 1 \end{bmatrix} \begin{bmatrix} 5 & -10 \\ 0 & 1 \end{bmatrix}$$

5. Find LU, Cholesky, QR decomposition for the matrix $A = \begin{bmatrix} 3 & 4 \\ 1 & 2 \end{bmatrix}$.

6. Find eigenvalue decompositions of the following matrices if possible:

$$A = \begin{bmatrix} 3 & 4 \\ 1 & 2 \end{bmatrix}, \quad B = \begin{bmatrix} 1 & 4 & 7 \\ 2 & 5 & 8 \\ 3 & 6 & 9 \end{bmatrix}.$$

7. Find a SVD of the matrix

$$A = \begin{bmatrix} 1 & 2 \\ 3 & 4 \\ 0 & 0 \\ 0 & 0 \end{bmatrix}.$$

8. Consider matrix $A = \begin{bmatrix} -2 & 11 \\ -10 & 5 \end{bmatrix}$. Determine a SVD. This decomposition is not unique, and find another SVD.

9. Suppose A is an $n \times n$ matrix over the complex numbers, and $A = S\Sigma V^*$ is a SVD. Find an eigenvalue decomposition of the Hermitian $\begin{bmatrix} 0 & A^* \\ A & 0 \end{bmatrix}$.

10. Find a Jordan decomposition of the matrix $A = \begin{bmatrix} 3 & 1 & 0 \\ -1 & 1 & 0 \\ 3 & 2 & 2 \end{bmatrix}$.

11. Write the matrix $A = \begin{bmatrix} -1 & -1 & 0 \\ 0 & -1 & 2 \\ 0 & 0 & -1 \end{bmatrix}$ in Jordan canonical form.

12. Apply Jacobi iterative method to solve

$$5x - 2y + 3z = -1$$
$$-3x + 9y + z = 2$$
$$2x - y - 7z = 3.$$

13. Apply Jacobi iterative method to solve

$$5x - 2y + 3z = -1$$
$$-3x + 9y + z = 2$$
$$2x - y - 7z = 3.$$

Continue iterations until two successive approximations are identical when rounded to three significant digits.

14. Apply the Gauss-Seidel iterative method to solve

$$5x - 2y + 3z = -1$$
$$-3x + 9y + z = 2$$
$$2x - y - 7z = 3.$$

Continue iterations until two successive approximations are identical when rounded to three significant digits.

15. Apply other methods to solve

$$\begin{aligned} 5x - 2y + 3z &= -1 \\ -3x + 9y + z &= 2 \\ 2x - y - 7z &= 3. \end{aligned}$$

BIBLIOGRAPHY

[CWZ04] R. M. Corless, S. M. Watt, and Lihong Zhi, *Qr factoring to compute the gcd of univariate approximate polynomials*, IEEE Transactions on Signal Processing **52** (2004), no. 12, 3394–3402.

[HJ85] Roger A. Horn and Charles R. Johnson, *Matrix analysis*, Cambridge University Press, Boston, United States, 1985.

[Lai69] M. A. Laidacker, *Another theorem relating sylvester's matrix and the greatest common divisor*, Mathematics Magazine **42** (1969), no. 3, 126–128.

[Map16] Maplesoft, *Maple*, a division of Waterloo Maple Inc., Waterloo, Ontario, 2016.

[R C13] R Core Team, *R: A language and environment for statistical computing*, R Foundation for Statistical Computing, Vienna, Austria, 2013.

[SJ05] William Stein and David Joyner, *SAGE: System for algebra and geometry experimentation*, Communications in Computer Algebra (SIGSAM Bulletin) (July 2005), http://www.sagemath.org.

[Tiw04] Ashish Tiwari, *Termination of linear programs*, Computer Aided Verification: 16th International Conference, CAV 2004, Boston, MA, USA, July 13–17, 2004. Proceedings (Rajeev Alur and Doron A. Peled, eds.), Springer-Verlag, Berlin Heidelberg, 2004, pp. 70–82.

[TM] The MathWorks, Inc. *Matlab 8.0 and statistics toolbox 8.1*, Natick, Massachusetts, United States.

[Wol15] Wolfram, *Mathematica, 10.3 ed.*, Wolfram Research, Inc., Champaign, Illinois, 2015, https://www.wolfram.com.

[WWK01] Wenping Wang, Jiaye Wang, and Myung-Soo Kim, *An algebraic condition for the separation of two ellipsoids*, Comput. Aided Geom. Des. 18 (2001), no. 6, 531–539.

INDEX

www.ingramcontent.com/pod-product-compliance
Lightning Source LLC
Chambersburg PA
CBHW061921190326
41458CB00009B/2619
9781683923763